An Introduction to
Fast Fourier Transform
Methods for Partial
Differential Equations,
with Applications

ELECTRONIC & ELECTRICAL ENGINEERING RESEARCH STUDIES

APPLIED AND ENGINEERING MATHEMATICS SERIES

Series Editor: **Professor P. C. Kendall,** *Department of Electronic and Electrical Engineering, The University of Sheffield, England*

An Introduction to Fast Fourier Transform Methods for Partial Differential Equations, with Applications

Morgan Pickering

University of Sheffield, England

RESEARCH STUDIES PRESS
Letchworth, Hertfordshire, England

JOHN WILEY & SONS INC.
New York · Chichester · Toronto · Brisbane · Singapore

RESEARCH STUDIES PRESS LTD
58B Station Road, Letchworth, Herts. SG6 3BE, England

Marketing and Distribution
Australia, New Zealand, South-east Asia:
Jacaranda-Wiley Ltd., Jacaranda Press
JOHN WILEY & SONS INC.
GPO Box 859, Brisbane, Queensland, 4001, Australia

Canada:
JOHN WILEY & SONS CANADA LIMITED
22 Worcester Road, Rexdale, Ontario, Canada

Europe, Africa:
JOHN WILEY & SONS LIMITED
Baffins Lane, Chichester, West Sussex, England

North and South America and the rest of the world:
JOHN WILEY & SONS INC.
605 Third Avenue, New York, NY 10158, USA

Library of Congress Cataloging-in-Publication Data

Pickering, Morgan, 1943–
    An introduction to fast fourier transform methods
for partial differential equations, with applications.

    (Electronic & electrical engineering research
studies. Applied and engineering mathematics series; 4)
    Bibliography: p.
    Includes index.
    1. Fourier transformations.   2. Differential
equations, Partial—Numerical solutions.   I. Title.
II. Series.
QA403.5.P53   1986   515.3'53   86-17891
ISBN 0 471 91261 1 (Wiley)

British Library Cataloguing in Publication Data

Pickering, Morgan
    An introduction to fast Fourier transform methods for
partial differential equations, with applications.
    —(Electronic and electrical engineering research
studies. Applied and engineering mathematics series;
4)
    1. Differential equations, Partial—Numerical
solutions   2. Fourier transformations
    I. Title   II. Series
    515.3'53   QA374

    ISBN 0 86380 045 9
    ISBN 0 471 91261 1 Wiley

    ISBN 0 86380 045 9 (Research Studies Press Ltd.)
    ISBN 0 471 91261 1 (John Wiley & Sons Inc.)

Printed in Great Britain by Galliard (Printers) Ltd, Great Yarmouth

Preface

Over the last twenty years there have been significant developments in numerical methods for the solution of partial differential equations, particularly in the field of rapid elliptic solvers. With the vast increase in available computing power, especially in the last decade, it can be argued that any numerical method which "works" will suffice to solve a given problem. This view may have some validity for small problems but, for large-scale scientific and engineering computations, suitable fast solvers are considerably more effective than earlier techniques.

For the newcomer the specialised nature of most fast solver algorithms can be unattractive and this may be true also for experienced researchers whose primary interests lie in areas other than numerical analysis. Of course, as with any unfamiliar topic, a certain level of investment of time and energy is necessary to get started, but the experience of many workers, including that of the author, verifies that in this particular area, the returns are well worth the effort. In fact some problems are insoluble without such an investment of effort. One of my hopes is that this book will help research students and research workers who wish to make use of modern developments in numerical methods. In a volume of this size it is not possible to cover all the detailed variations of the available algorithms and, with this in mind, the book

concentrates particularly on Fast Fourier Transform (FFT) methods
and on what can be regarded as standard methodology. Virtually all
the material covered can be accessed through journals and other
books; where a topic is highly complicated or specialised, suitable
references are given for further reading.

It is unfortunate that in many undergraduate science and
engineering courses there is usually insufficient time available to
cover adequately even the essential ideas of relatively recent
developments in numerical techniques. I am convinced, however,
that a modern final year honours course on the numerical solution
of partial differential equations should include some discussion of
fast solvers. For this reason a few lectures, based largely on the
introductory FFT material covered in Chapter 1, have been included
in such a course at Sheffield for a number of years. Chapter 2
concentrates on the mathematics of the transforms used in solving
suitable partial differential equations although the point should
be made here that much of the programming effort required for these
algorithms has been removed by the availability of a variety of
packages which, inter alia, efficiently implement the complex FFT
or the FFT of real data. A Fortran program for the complex FFT is
given in Appendix II and is reproduced from Brigham's book "The
Fast Fourier Transform" which also contains many references to the
FFT programs of other authors. The Computer Physics Communications
program library, Queen's University, Belfast, is also a useful
source of several suitable packages, as is the NAG scientific

software library. Chapter 3 examines various problems which can be solved directly by the FFT method and Chapter 4 discusses the basic ideas of the cyclic reduction method and the FACR algorithm which is a combination of the methods of Fourier analysis and cyclic reduction. Chapter 5 considers the application of fast solvers to problems on irregular regions using the so-called capacity matrix approach. The iterative use of fast solvers for solving more complicated equations and the use of a numerical Laplace transform technique for the solution of certain time-dependent problems are discussed in Chapter 6.

I am pleased to thank Peter Kendall for his invitation to write this book and for his encouragement and helpful comments throughout the period that the writing was undertaken. Furthermore, it is a pleasure to acknowledge David Burley's constructive criticism of an earlier draft. I am also very grateful to Suzanne Bedford, Madeleine Floy, Christina Vickers and Pamela Holloway for their untiring assistance at all levels in the production of the camera-ready copy.

Sheffield

May 1986   Morgan Pickering

Contents

Chapter 3.   FFT Solution of Partial Differential Equations

Chapter 4.   Cyclic Reduction

Chapter 5. Irregular Regions

Chapter 6. Two methods for more general problems

Appendix 1

CHAPTER 1
Basic Preliminaries

1.1 Introduction

In this Chapter we introduce some of the theory and
methodology associated with the numerical solution of partial
differential equations using FFT techniques. One general feature
of such problems is that the numerical method employed is basically
"matrix decomposition" [see, for example, Buzbee et al (1970)]
which is essentially the discrete equivalent of the method of
separation of variables.

For a particular class of problems advantage can be taken of
FFT methods at certain stages in the overall computational
procedure and an illustrative example is formulated and discussed
in Sections 1.2 and 1.3. Section 1.4 outlines some of the basic
ideas of FFT computation and Section 1.5 describes some suitable
algorithms for solving the systems of linear equations generated by
the method. Detailed numerical results for a particular problem
are presented in Section 1.6.

1

1.2 Poisson's equation in a rectangle

In order to illustrate the basic features of the method of discrete Fourier analysis as applied to the numerical solution of partial differential equations, we consider solving Poisson's equation

$$\frac{\partial^2 \phi}{\partial x^2} + \frac{\partial^2 \phi}{\partial y^2} = q(x,y) \ , \qquad (1.1)$$

over a rectangular region $0 \leqslant x \leqslant L_x$, $0 \leqslant y \leqslant L_y$ with Dirichlet boundary conditions,

$$\phi(0,y)=f(y), \quad \phi(L_x,y)=F(y), \qquad (1.2)$$

$$\phi(x,0)=g(x), \quad \phi(x,L_y)=G(x).$$

At a grid point $(x_j, y_s) = (jh_x, sh_y)$, $j=1,2,\ldots,R$; $s=1,2,\ldots,S$; where $h_x = L_x/(R+1)$ and $h_y = L_y/(S+1)$, we approximate equation (1.1) by

$$\phi_{j-1,s} - 2(1+h^2)\phi_{j,s} + \phi_{j+1,s} + h^2(\phi_{j,s+1} + \phi_{j,s-1}) = h_x^2 q_{j,s}, \qquad (1.3)$$

where $h = h_x/h_y$ and the usual second-order central-difference approximations to derivatives have been employed, for example

$$\frac{\partial^2 \phi}{\partial x^2} \approx \frac{(\phi_{j-1,s} - 2\phi_{j,s} + \phi_{j+1,s})}{h_x^2}.$$

Thus for any row $s=1,2,\ldots,S$ of our grid equations (1.3) may be written as

$$-2(1+h^2)\phi_{1,s} + \phi_{2,s} + h^2(\phi_{1,s+1} + \phi_{1,s-1}) \qquad\qquad = h_x^2 q_{1,s} - f_s$$

$$\cdots\cdots\cdots\cdots\cdots\cdots\cdots\cdots\cdots\cdots\cdots\cdots\cdots\cdots\cdots \qquad\qquad \cdots$$

$$\phi_{j-1,s} - 2(1+h^2)\phi_{j,s} + \phi_{j+1,s} + h^2(\phi_{j,s+1} + \phi_{j,s-1}) = h_x^2 q_{j,s} \qquad (1.4)$$

$$\cdots\cdots\cdots\cdots\cdots\cdots\cdots\cdots\cdots\cdots\cdots\cdots\cdots\cdots\cdots \qquad\qquad \cdots$$

$$\phi_{R-1,s} - 2(1+h^2)\phi_{R,s} + h^2(\phi_{R,s+1} + \phi_{R,s-1}) = h_x^2 q_{R,s} - F_s,$$

where the known boundary values of ϕ on $x=0,L_x$ have been incorporated into the right-hand sides of the first and last of equations (1.4) in the usual way and $f_s=f(y_s)$, $F_s=F(y_s)$. In order to express the totality of S sets of equations (1.4) in a compact form we introduce the notation

$$\underset{\sim}{\phi}_s=(\phi_{1,s},\phi_{2,s},\cdots,\phi_{R,s})^T \tag{1.5}$$

and

$$\underset{\sim}{Q}_s=(h_x^2q_{1,s}-f_s,h_x^2q_{2,s},\cdots,h_x^2q_{R-1,s},h_x^2q_{R,s}-F_s)^T \tag{1.6}$$

so that

$$A\underset{\sim}{\phi}_1+h^2I\underset{\sim}{\phi}_2 \qquad\qquad = \underset{\sim}{Q}_1^+$$

$$\cdots\cdots\cdots \qquad\qquad\qquad \cdot$$

$$h^2I\underset{\sim}{\phi}_{s-1}+A\underset{\sim}{\phi}_s+h^2I\underset{\sim}{\phi}_{s+1} \qquad = \underset{\sim}{Q}_s^+ \tag{1.7}$$

$$\cdots\cdots\cdots\cdots \qquad\qquad \cdot$$

$$h^2I\underset{\sim}{\phi}_{S-1}+A\underset{\sim}{\phi}_S = \underset{\sim}{Q}_S^+,$$

where I denotes the unit matrix of order R, A is an R-th order symmetric tridiagonal matrix given by

$$A = \begin{bmatrix} -2(1+h^2) & 1 & & & \\ 1 & -2(1+h^2) & 1 & & \\ & & \cdots\cdots\cdots & & \\ & & \cdots\cdots\cdots & & \\ & & & 1 & -2(1+h^2) & 1 \\ & & & & 1 & -2(1+h^2) \end{bmatrix}, \tag{1.8}$$

$$Q_1^+ = Q_1 - h^2 I \psi_0,$$

$$Q_s^+ = Q_s, \qquad (s=2,3,\ldots,S-1)$$

$$Q_S^+ = Q_S - h^2 I \psi_{S+1}$$

(1.9)

and ψ_0, ψ_{S+1} denote vectors whose components are known values of ϕ on $y=0$ and $y=L_y$, respectively,

$$\psi_0 = (g_1, g_2, \ldots, g_R)^T$$

and $\qquad \psi_{S+1} = (G_1, G_2, \ldots, G_R)^T,$

(1.10)

where $g_j = g(x_j)$ and $G_j = G(x_j)$.

We expand each ψ_s and Q_s^+ in terms of the known eigenvectors x_r of the matrix A (see Appendix 1). Thus we set

$$\psi_s = \sum_{r=1}^{R} c_{r,s} x_r$$

(1.11)

and

$$Q_s^+ = \sum_{r=1}^{R} d_{r,s} x_r,$$

(1.12)

for $s=1,2,\ldots,S$, where

$$x_r = (\sin\theta_r, \sin2\theta_r, \ldots, \sin R\theta_r)^T,$$

$$\theta_r = \frac{r\pi}{R+1}$$

$(r=1,2,\ldots,R)$ (1.13)

and the eigenvectors satisfy the orthogonality relation

$$x_\zeta^T x_\eta = \frac{1}{2}(R+1)\delta_{\zeta\eta},$$

(1.14)

where $\delta_{\zeta\eta}$ denotes the Kronecker delta function. Substituting

equations (1.11) and (1.12) into (1.7) and making use of (1.14) we
obtain the tridiagonal systems

$$
\left.\begin{array}{rl}
\lambda_r c_{r,1} + h^2 c_{r,2} & = d_{r,1} \\
\cdots\cdots\cdots\cdots & \quad . \\
h^2 c_{r,s-1} + \lambda_r c_{r,s} + h^2 c_{r,s+1} & = d_{r,s} \\
\cdots\cdots\cdots\cdots\cdots & \quad . \\
h^2 c_{r,S-1} + \lambda_r c_{r,S} & = d_{r,S},
\end{array}\right\} (r=1,2,\ldots,R) \qquad (1.15)
$$

where λ_r denotes an eigenvalue of A given by

$$
\lambda_r = -2(1+h^2) + 2\cos\frac{r\pi}{R+1} \qquad (1.16)
$$

and

$$
d_{r,s} = \frac{\underset{\sim}{x}_r^T \underset{\sim}{Q}_s^+}{|\underset{\sim}{x}_r|^2} = \frac{2}{R+1} \sum_{j=1}^{R} Q_{j,s}^+ \sin j\theta_r . \qquad (1.17)
$$

Equations (1.15) denote R independent tridiagonal systems of
equations of order S for the Fourier harmonics $c_{r,s}$ (r=1,2,...,R;
s=1,2,...,S) and we note that the decoupling of system (1.7) has
been accomplished directly as a result of the orthogonality of the
eigenvectors $\underset{\sim}{x}_r$. Once the systems (1.15) have been solved for $c_{r,s}$
the values of $\phi_{j,s}$ can be reconstructed (or 'synthesised') using
equation (1.11), which in component form may be written as

$$
\phi_{j,s} = \sum_{r=1}^{R} c_{r,s} \sin j\theta_r . \qquad (1.18)
$$

The overall procedure may be summarised as follows:

(i)   Calculate $Q_{j,s}$ for $j=1,2,\ldots,R$; $s=1,2,\ldots,S$.

(ii)   Compute $d_{r,s}$ from (1.17); (overwriting $Q_{j,s}^{+}$).

(iii)   Solve (1.15) for $c_{r,s}$ (overwriting $d_{r,s}$).

(iv)   Use (1.18) to synthesise $\phi_{j,s}$ (overwriting $c_{r,s}$).

1.3 Discussion

At first sight, the method does not appear to be computationally attractive since the evaluation of the Fourier sums directly using (1.17) and (1.18) is relatively time consuming. This can be seen if we estimate the number of computer operations required to perform the calculation, where, by 'operation' we mean a multiplication and the addition which normally accompanies it in such summations. In order to simplify the discussion we concentrate on the case R=S=N.

The calculation of $d_{r,s}$ requires essentially N^2 operations for each s, and similarly each Fourier synthesis via (1.18) requires N^2 operations. The N sets of tridiagonal equations can each be solved efficiently in the order of N operations (see Section 1.5) giving a total operation count for this stage which is proportional to N^2. Hockney (1965) employed a method described by Whittaker & Robinson (1944), which allows the computation of Fourier analysis or synthesis in approximately $N^2/36$ operations, where $N=12 \times 2^p$ and p is an integer $\geqslant 1$. The methods of Cooley and Tukey (1965), Gentleman and Sande (1966) achieve a Fourier analysis or synthesis in an operations count proportional to $Nlog_2N$, where usually $N=2^P$ and similar efficiency is achieved in a method described by Hockney (1970). Precise operation counts for a given problem depend particularly on the boundary conditions. Hockney's method for Poisson's equation with Dirichlet or Neumann conditions on $x=0,L_x$ requires approximately twice the work of the corresponding periodic problem. Cooley et al (1970) have shown how the Fourier analysis and synthesis can be organised so that the computation for any of the above boundary conditions takes much the same time as Hockney's periodic case and we describe their algorithms in some detail in the following Chapter.

In order to reduce the amount of Fourier analysis and synthesis required by the method, block eliminations may be performed on the original set of N^2 linear equations (1.7) to produce a rather more complicated set of equations which involve only the unknowns on the even numbered lines of the grid. These equations also may be solved by the Fourier method. A combination of odd/even reduction and Fourier analysis forms the basis of the highly efficient method of Fourier analysis and cyclic reduction FACR(ℓ), described in detail, for example, by Hockney (1970), Swartztrauber (1977) and Temperton (1980), in which ℓ levels of reduction are carried out before Fourier analysis is performed. We consider such an approach in Chapter 4.

A detailed comparison of operation counts for several direct and iterative methods for the discrete Poisson equation on a rectangle has been given by a number of authors, notably Dorr (1970), Hockney (1970), Swartztrauber (1977) and Temperton (1980). In particular Swartztrauber compares cyclic reduction, Fourier analysis and FACR(ℓ) and shows that the operation count for the FACR(ℓ) method is $O(N^2 \log_2 \log_2 N)$ for a 'best' choice of $\ell = \ell^* \simeq \log_2 \log_2 N - 1$ ($\ell = 2$ or 3 for practical meshes) whereas if either Fourier analysis or cyclic reduction is used independently, the corresponding operation count is $O(N^2 \log_2 N)$. In practice, for values of N between 63 and 1023, Swartztrauber obtained reductions in program execution time varying between about 8 and 55 per cent. using the FACR(ℓ^*) method as compared with cyclic reduction or Fourier analysis alone. Some perspective can be gained from the fact that for N=63 he found that the program execution time was roughly the same as that for 5 iterations of the successive over-relaxation method applied to the same problem.

1.4   Summation of complex finite Fourier series

A method which efficiently evaluates the sum of a complex finite Fourier series as described by Cooley and Tukey(1965) is based on the methods of Good(1958) and earlier workers. Essentially their technique relies on the number of terms, n, being factorisable. As an introduction to algorithms of this type we illustrate the procedure for $n = m_1 m_2$ and explain how such a procedure could be used to compute (1.17) or (1.18). A more general case is considered in detail in Chapter 2.

We consider the problem of evaluating the Fourier series

$$X(j) = \sum_{r=0}^{n-1} A(r)W_n^{jr}, \quad (j=0,1,\ldots,n-1) \qquad (1.19)$$

where the Fourier harmonics $A(r)$ are in general complex,

$W_n = \exp(2\pi i/n)$ and $i = \sqrt{-1}$. Equation (1.19) is also known as the

inverse discrete Fourier transform (IDFT) since the relation for

$A(r)$ in terms of $X(j)$

$$A(r) = \frac{1}{n} \sum_{j=0}^{n-1} X(j)W_n^{-jr}, \quad (r=0,1,\ldots,n-1) \qquad (1.20)$$

is usually called the discrete Fourier transform (DFT). This

terminology follows that for the corresponding continuous case

where it is well known that, for a periodic function, the

coefficients in the usual Fourier series and those derived by

means of the Fourier integral are the same (see, for example,

Brigham (1974),p78). Relation (1.20) may be derived easily from

(1.19) by noting that

$$\sum_{j=0}^{n-1} W_n^{j(r-\bar{r})} = n\delta_{r\bar{r}} . \qquad (1.21)$$

In some definitions the factor $1/n$ may appear in (1.19) rather than

(1.20) and some authors use a factor $1/\sqrt{n}$ in both.

An important point to note is that (1.19) may be written as

$$X(j) = \Big[\sum_{r=0}^{n-1} A^*(r)W_n^{-jr} \Big]^* , \qquad (1.22)$$

where * denotes complex conjugate. This result demonstrates that
any method developed to evaluate (1.19) may be used for (1.20) by
replacing $A(r)$ by $X^*(j)/n$ as input and taking the complex conjugate
of the result. We concentrate here on (1.19).

The indices r and j are written in the form

$$r = r_1 m_2 + r_0, \quad r_0 = 0, 1, \ldots, m_2 - 1, \quad r_1 = 0, 1, \ldots, m_1 - 1,$$

$$j = j_1 m_1 + j_0, \quad j_0 = 0, 1, \ldots, m_1 - 1, \quad j_1 = 0, 1, \ldots, m_2 - 1,$$

so that (1.19) may be expressed as

$$X(j_1, j_0) = \sum_{r_0} \sum_{r_1} A(r_1, r_0) W_n^{j r_1 m_2} W_n^{j r_0}. \tag{1.23}$$

Now

$$W_n^{j r_1 m_2} = W_n^{(j_1 m_1 + j_0) r_1 m_2} = W_n^{j_1 r_1 n} W_n^{j_0 r_1 m_2} = W_n^{j_0 r_1 m_2}, \tag{1.24}$$

and thus the inner sum of (1.23) depends only on j_0 and r_0 and can
be defined as

$$A_1(j_0, r_0) = \sum_{r_1} A(r_1, r_0) W_n^{j_0 r_1 m_2}$$

$$\tag{1.25}$$

so that $\quad X(j_1, j_0) = \sum_{r_0} A_1(j_0, r_0) W_n^{(j_1 m_1 + j_0) r_0}.$

An approximate operations count is as follows. There are n
elements of A_1 each requiring m_1 operations giving a total of nm_1
operations to obtain A_1. Similarly nm_2 operations are required to
obtain X from A_1, giving an overall total

$$T = n(m_1 + m_2) \tag{1.26}$$

operations.

For a more general factorisation

$$n = m_1 m_2 \cdots m_k, \qquad (1.27)$$

it is easy to see that

$$T = n(m_1 + m_2 + \ldots + m_k) \qquad (1.28)$$

and that using as many factors as possible provides a minimum to (1.28) (factors of 2 can be combined in pairs without loss). In particular if $m_1 = m_2 = \ldots = m_k = m$ then $n = m^k$ and

$$T = nmk = nm\log_m n = \frac{m}{\log_2 m} (n\log_2 n). \qquad (1.29)$$

Some values of $m/\log_2 m$ are given in Table 1.1 and show clearly that m=3 is the most efficient. m=2 or 4 are usually used, however, since they have other advantages.

m	$m/\log_2 m$
2	2.00
3	1.88
4	2.00
5	2.15
6	2.31
7	2.49

Table 1.1   Values of m and $m/\log_2 m$.

We leave the details of one such algorithm to Chapter 2 and there give a more precise count of operations.

In addition to being computationally efficient FFT methods are also numerically stable. The propagation of round-off errors for such methods was studied, for example, by Gentleman and Sande (1966) who derived an upper bound for $R^{(e)}$, the ratio of the Euclidean norm of the error to the Euclidean norm of the data sequence (the Euclidean norm of a sequence is the square root of the sum of squares of its elements). They found that for $n = m^k$

$$R^{(e)}/R^{(d)} = k/m^{3(k-1)/2}, \qquad (1.30)$$

where $R^{(d)}$ denotes the ratio for the direct calculation. Some values of $R^{(e)}/R^{(d)}$ for $m=2$ are given in Table 1.2 and the results of computational experiments for a number of random data sequences of various lengths are reported in Gentleman and Sande's paper.

k	$k/2^{3(k-1)/2}$
2	0.707
3	0.375
4	0.177
5	0.078
6	0.033
7	0.014

Table 1.2 Values of k and $R^{(e)}/R^{(d)}$ for $m=2$.

Automatic round-off error analysis in discrete linear transforms has been studied by Bois and Vignes (1982).

Any algorithm for the evaluation of (1.19) can be used to compute the sine transform required by equation (1.17) by a method analogous to that for the usual half-range Fourier sine series for continuous data. With n replaced by 2n equation (1.19) is of the form

$$X(j) = \sum_{r=0}^{2n-1} A(r) \left[\cos \frac{\pi j r}{n} + i \sin \frac{\pi j r}{n} \right] \quad (j=0,1,\ldots,2n-1) \quad (1.31)$$

where we now assume that $A(r)$ is real, so that

$$X_R(j) = \sum_{r=0}^{2n-1} A(r) \cos \frac{\pi j r}{n} \quad (1.32)$$

and

$$X_I(j) = \sum_{r=0}^{2n-1} A(r) \sin \frac{\pi j r}{n}, \quad (1.33)$$

where $X_R(j)$ and $X_I(j)$ denote, respectively, the real and imaginary parts of $X(j)$. We can ensure that $X_R(j)=0$ by constructing the data so that

$$A(0) = A(n) = 0 \quad (1.34)$$

and

$$A(\mu) = -A(2n-\mu), \quad (\mu=1,2,\ldots,n-1) \quad (1.35)$$

so that

$$X_I(j) = 2 \sum_{r=1}^{n-1} A(r) \sin \frac{\pi j r}{n}, \quad (1.36)$$

which is of the form (1.18) with

$$R = n-1. \quad (1.37)$$

Equation (1.17) also may be computed using this kind of technique.

A similar approach may be used to obtain the cosine transform, which arises from problems with Neumann conditions on $x=0,L_x$. More efficient techniques for calculating both the sine transform and cosine transform will be presented in Chapter 2, together with efficient methods for determining the appropriate transforms for problems with periodic boundary conditions and Dirichlet-Neumann conditions.

1.5   The solution of tridiagonal systems

We consider an N-th order tridiagonal system of the form

$$b_1 x_1 + c_1 x_2 \qquad\qquad\qquad = d_1$$
$$a_2 x_1 + b_2 x_2 + c_2 x_3 \qquad\qquad = d_2$$
$$\dots\dots\dots\dots\dots\dots\dots\dots\dots\dots\dots\dots \qquad\qquad . \qquad\qquad (1.38)$$
$$a_{N-1} x_{N-2} + b_{N-1} x_{N-1} + c_{N-1} x_N = d_{N-1}$$
$$a_N x_{N-1} + b_N x_N = d_N.$$

Using Gaussian elimination to reduce the system to a simple upper-triangular form and solving these equations by back substitution we may derive the following algorithm for the solution of (1.38).

Form the quantities

$$\left.\begin{aligned} p_i &= a_i q_{i-1} + b_i, \qquad (q_0=0)\\ q_i &= -c_i/p_i \\ \text{and } \quad u_i &= (d_i - a_i u_{i-1})/p_i, \qquad (u_0=0) \end{aligned}\right] \quad (i=1,2,\dots,N) \qquad (1.39)$$

set $x_N = u_N$

and calculate

$$x_i = q_i x_{i+1} + u_i. \quad (i=N-1, N-2, \ldots, 1) \tag{1.40}$$

It is clear that p_i, q_i and u_i may overwrite, respectively b_i, c_i and d_i and that x_i may overwrite d_i. A count of operations shows that this algorithm requires approximately 5N multiplications, 3N additions and N divisions or 3N multiplications and 2N additions per right-hand side if precomputed coefficients are used.

The above method is often called the Thomas algorithm [Thomas (1949)] although it seems likely that it first appeared in published form in a paper by Bruce et al (1953). Wilkinson (1961) has shown that this type of method is stable with respect to the growth of round-off errors if the coefficient matrix is diagonally dominant. For our system (1.38) this implies that we must have $|b_i| > |a_i| + |c_i|$, $i=2,3,\ldots,N-1$ and $|b_1| > |c_1|$, $|b_N| > |a_N|$.

Problems with Neumann conditions also give rise to tridiagonal systems as illustrated by the example of Section 3.5.2 in Chapter 3, but if periodic conditions are imposed the systems corresponding to (1.15) are no longer simply tridiagonal but are of the cyclic form

$$
\begin{aligned}
b_1 x_1 + c_1 x_2 + & & + a_1 x_N &= d_1 \\
a_2 x_1 + b_2 x_2 + c_2 x_3 & & &= d_2 \\
& \cdots\cdots\cdots\cdots\cdots\cdots\cdots\cdots\cdots\cdots & & \\
& a_{N-1} x_{N-2} + b_{N-1} x_{N-1} + c_{N-1} x_N &= d_{N-1} \\
c_N x_1 + & & + a_N x_{N-1} + b_N x_N &= d_N.
\end{aligned}
\tag{1.41}
$$

Probably what is regarded as the 'standard' algorithm for solving (1.41) may be found, for example, in Ahlberg et al (1967, p15) and is based on the bordering method for finding an inverse matrix. Defining M to be the tridiagonal matrix of order $N-1$ obtained by deleting the last row and column of the matrix of coefficients of (1.41) and defining

$$\left.\begin{aligned}
\bar{x} &= (x_1, x_2, \ldots, x_{N-1})^T \\
\bar{d} &= (d_1, d_2, \ldots, d_{N-1})^T \\
g_1 &= (a_1, 0, 0, \ldots, 0, c_{N-1})^T \\
\text{and} \quad g_2 &= (c_N, 0, 0, \ldots, 0, a_N)^T
\end{aligned}\right\} \qquad (1.42)$$

we see that system (1.41) may be written in the form

$$\left.\begin{aligned}
M\bar{x} + g_1 x_N &= \bar{d} \\
g_2^T \bar{x} + b_N x_N &= d_N
\end{aligned}\right\} . \qquad (1.43)$$

Writing $\underset{\sim}{v} = M^{-1}\bar{d}$ we obtain

$$\bar{x} = \underset{\sim}{v} - M^{-1} g_1 x_N \qquad (1.44)$$

and

$$g_2^T \underset{\sim}{v} + \left[b_N - g_2^T M^{-1} g_1 \right] x_N = d_N \qquad (1.45)$$

from which it is clear that once x_N is determined from (1.45), the components of $\bar{x}$ may be calculated using (1.44). Thus, making use of the tridiagonal algorithm already described, equations (1.41) may be solved by the following procedure.

Form the quantities

$$p_i = a_i q_{i-1} + b_i, \qquad (q_0 = 0)$$

$$q_i = -c_i/p_i,$$

$$u_i = (d_i - a_i u_{i-1})/p_i, \quad (u_0 = 0) \qquad (i=1,2,\ldots N-1) \qquad (1.46)$$

$$s_i = -a_i s_{i-1}/p_i \qquad (s_0 = 1)$$

and

$$t_i = q_i t_{i+1} + s_i \qquad (t_N = 1)$$
$$\qquad\qquad\qquad\qquad\qquad\qquad (i=N-1,N-2,\ldots,1) \qquad (1.47)$$
$$v_i = q_i v_{i+1} + u_i. \qquad (v_N = 0)$$

Solve

$$(c_N t_1 + a_N t_{N-1} + b_N)x_N = d_N - c_N v_1 - a_N v_{N-1} \qquad (1.48)$$

for x_N and determine $x_{N-1}, x_{N-2}, \ldots, x_1$ from

$$x_i = t_i x_N + v_i. \qquad (i=N-1,N-2,\ldots 1) \qquad (1.49)$$

The storage can be arranged in a similar manner to the tridiagonal case except that s_i may overwrite a_i, t_i and v_i may overwrite respectively s_i and u_i and finally x_i may overwrite v_i. The algorithm requires approximately 9N multiplications, 5N additions and N divisions, or 4N multiplications and 3N additions per right-hand side if precomputed coefficients are used. We note here that, for example, for the two-dimensional Poisson problem in a rectangle with periodic conditions in both the x and y directions the discrete problem is singular and a procedure similar to that outlined in Section 3.5.2, for a purely Neumann problem, may be utilised to obtain a solution.

Other methods which might be used to solve (1.41) include those of Evans & Atkinson (1970) who applied Gaussian elimination directly to (1.41), Tang (1969), who employed an elimination method with the super-diagonal elements used as pivots, and Pickering (1984), who formulated a method which makes use of the solution of the associated tridiagonal system (1.38). All these methods require rather more arithmetic than the above algorithm.

Several highly efficient algorithms have been devised for specialised coefficient matrices of the form (1.38) or (1.41). In particular Evans (1971, 1972) described a factorisation method for particular symmetric tridiagonal and circulant matrices, whereas Hockney (1965, 1970) used a method based on cyclic reduction which, for Poisson's equation with periodic conditions, can solve the equations for the Fourier harmonics in approximately the same number of operations as the tridiagonal case. Temperton (1975) described a method which relies on an a priori knowledge of x_1 and presented some comparative results using various algorithms.

1.6   A numerical example

Numerical solutions of equation (1.1) over the unit square $0 \leqslant x, y \leqslant 1$ with

$$q(x,y) = (x^2 + y^2)e^{xy}, \qquad (1.50)$$

$$f(y) = g(x) = 1, \qquad (1.51)$$

and
$$F(y) = e^y, \quad G(x) = e^x \qquad (1.52)$$

were determined for various square grid sizes using the methods outlined in the preceding Sections.   The program was written in Fortran and run on an ICL 1906S machine and the discrete Fourier series was summed using a standard library FFT package based on the Cooley-Tukey algorithm.

The analytic solution of (1.1) with q(x,y) given by (1.50) and boundary conditions (1.51) and (1.52) is easily shown to be

$$\phi(x,y) = e^{xy}. \qquad (1.53)$$

The numerically obtained solutions were compared with the above formula for each of the grid sizes used and values of the maximum modulus error and R.M.S. error thus determined are given in Table 1.3.

N	Max. modulus error	R.M.S. error	Max. modulus residual	Execution time (sec)
7	4.6,−5	3.8,−6	1.5,−10	0.06
15	1.2,−5	4.3,−7	2.6,−10	0.21
31	3.1,−6	5.1,−8	2.9,−10	0.73
63	7.6,−7	6.1,−9	5.2,−10	2.93
127	1.8,−7	7.0,−10	6.9,−10	11.80

Table 1.3   Results for test problem.   The numbers are in standard floating point form, e.g.   4.6,−5 means 4.6×10^{-5}.

The maximum modulus error was found to occur near to $(x,y)=(1,1)$ in all cases and shows clearly the second-order nature of our approximation, since for each halving of the mesh size this error is reduced by roughly a factor of four.   The R.M.S. error is reduced by approximately an order of magnitude for each halving of the mesh size and is of the order of the maximum modulus residual for the largest value of N used.   The values of the maximum modulus residual shown in Table 1.3 were determined by substituting computed $\phi_{j,s}$ values into the appropriate finite-difference equations and, as expected, increase as N increases.

The execution times shown in Table 1.3 were determined by running the program many times for each value of N.   Typically, for the smaller values of N, the times given represent a mean of 300 complete runs of the program whereas, for the larger N values, 30 complete runs were performed.   For N sufficiently large we might expect Swartztrauber's asymptotic formula for arithmetic operations (Section 1.3) to correlate reasonably well with the measured times.   The ratio of execution times for N=127 and N=63 is 4.03 which may be compared with the value of 4 derived from the asymptotic formula by letting N tend to infinity.   This value is also in close agreement with the ratio of times for N=63 and N=31 but, in general, Table 1.3 indicates that the ratio of execution times for successive pairs of values of N decreases as N decreases, whereas the corresponding values of the ratio predicted by the formula increase.   For our particular problem some time is initially spent on calculating the exponentials which occur in (1.50) and (1.52). Taking this into account gives a 'corrected' ratio of execution times for N=127 and N=63 of 4.06.

Many factors of course affect program execution time. For example, time is spent on decision instructions, accessing array elements and general 'housekeeping' by the machine. Hockney (1978) has made a study of the performance of a number of Poisson solvers on various machines and concludes that particular methods perform best for a specific computer/compiler combination and that the operation count may not necessarily be the most important feature in estimating program execution time on some computers.

CHAPTER 2
Algorithms

2.1 Introduction

This Chapter is devoted almost entirely to a discussion of algorithms which are useful for solving partial differential equations using the discrete Fourier transform. In particular we describe in some detail a complex fast Fourier transform algorithm for $n=2^k$ and also the use of such an algorithm for the efficient evaluation of sine and cosine series for real data. These procedures, together with that for the real periodic transform, are those given by Cooley <u>et al</u> (1970) and Temperton (1980). Swartztrauber (1977) derived an efficient algorithm for the transform used in a symmetric Dirichlet problem by relating this transform to the periodic case. Details of this relation are given in Appendix 1 (Section A1.5).

We begin by describing a complex FFT algorithm for the very simple case $n=4=2^2$.

2.2 The case $n = 4 = 2^2$

We consider again relation (1.19)

$$X(j) = \sum_{r=0}^{n-1} A(r)W_n^{jr}, \quad (j=0,1,\ldots,n-1) \tag{2.1}$$

for computing the discrete complex Fourier series.   If we set

$$r = 2r_1 + r_0 \qquad r_0 = 0,1; \; r_1 = 0,1$$

and $\qquad j = 2j_1 + j_0 \qquad j_0 = 0,1; \; j_1 = 0,1$

it is clear that r_0 and r_1 are the binary digits of r and similarly

j_0 and j_1 are those of j.   Thus, writing A and X as functions of

the bits of their indices, (2.1) may be expressed as

$$X(j_1,j_0) = \sum_{r_0=0}^{1} \sum_{r_1=0}^{1} A(r_1,r_0)W_4^{j(2r_1+r_0)} \tag{2.2}$$

and since $W_4 = \exp(\pi i/2)$

$$W_4^{2jr_1} = W_4^{2(2j_1+j_0)r_1} = W_4^{2j_0r_1} \qquad ,$$

equations (2.2) may be written as

$$X(j_1,j_0) = \sum_{r_0=0}^{1} \left[\sum_{r_1=0}^{1} A(r_1,r_0)W_4^{2j_0r_1} \right] W_4^{(2j_1+j_0)r_0}. \tag{2.3}$$

By defining

$$A_1(j_0,r_0) = \sum_{r_1=0}^{1} A(r_1,r_0)W_4^{2j_0r_1} \tag{2.4}$$

and

$$A_2(j_0,j_1) = \sum_{r_0=0}^{1} A_1(j_0,r_0)W_4^{(2j_1+j_0)r_0} \tag{2.5}$$

we see that

$$X(j_1,j_0) = A_2(j_0,j_1). \tag{2.6}$$

Thus the final results obtained from the second (outer) sum are in bit-reversed order with respect to the required values $X(j_1, j_0)$. This natural consequence of the algorithm is usually called 'scrambling' of the results.

Some insight into computational details can be obtained by noting that in matrix form equations (2.4) and (2.5) may be written as

$$
\begin{bmatrix} A_1(0,0) \\ A_1(0,1) \\ A_1(1,0) \\ A_1(1,1) \end{bmatrix} = \begin{bmatrix} 1 & 0 & W_4^0 & 0 \\ 0 & 1 & 0 & W_4^0 \\ 1 & 0 & W_4^2 & 0 \\ 0 & 1 & 0 & W_4^2 \end{bmatrix} \begin{bmatrix} A(0,0) \\ A(0,1) \\ A(1,0) \\ A(1,1) \end{bmatrix}
\tag{2.7}
$$

and

$$
\begin{bmatrix} A_2(0,0) \\ A_2(0,1) \\ A_2(1,0) \\ A_2(1,1) \end{bmatrix} = \begin{bmatrix} 1 & W_4^0 & 0 & 0 \\ 0 & W_4^2 & 0 & 1 \\ 0 & 0 & 1 & W_4^1 \\ 0 & 0 & 1 & W_4^3 \end{bmatrix} \begin{bmatrix} A_1(0,0) \\ A_1(0,1) \\ A_1(1,0) \\ A_1(1,1) \end{bmatrix}
\tag{2.8}
$$

respectively. Thus the matrix multiplication of (2.1) has been decomposed into two matrix multiplications where each matrix contains several elements which are zero or 1. It should be noted that elements which arise from the first term of summations (2.4) or (2.5) $(r_0 = r_1 = 0)$ have been written as 1; other combinations of indices which give rise to W_4^0 in these equations have been left as W_4^0 in order to indicate the general form of the result. Thus, for example, the calculation of $A_1(1,0)$ requires one complex multiplication and one complex addition, and the other elements of A_1 and A_2 calculated via (2.7) and (2.8) each require the same number of operations. A further saving of arithmetic can be achieved by noting that both $A_1(0,0)$ and $A_1(1,0)$ are to be computed in terms of $A(0,0)$ and $A(1,0)$; the only essential difference between the two computations is that in one W_4^0 is to be calculated and in the other we need to evaluate W_4^2. Since $W_4^2 = -W_4^0$ we see that

$$\left.\begin{array}{l} A_1(0,0) = A(0,0) + W_4^0 A(1,0) \\[2mm] A_1(1,0) = A(0,0) - W_4^0 A(1,0). \end{array}\right] \qquad (2.9)$$

Similar remarks apply to the calculation of the pairs $\left[A_1(0,1),A_1(1,1)\right]$ and $\left[A_2(0,0),A_2(0,1)\right]$ and also to the pair $\left[A_2(1,0),A_2(1,1)\right]$, here noting that $W_4^3 = -W_4^1$. The general form in each case is indicated by (2.9) and this implies that we can reduce the overall number of multiplications by a factor of 2 since, for example, $W_4^0 A(1,0)$ can be temporarily stored and equations (2.9) evaluated in situ as $A(0,0)$ and $A(1,0)$ are not used elsewhere in the calculation. Thus the overall computation requires 4 multiplications and 8 additions as compared with 16 multiplications and 12 additions if (2.1) were evaluated directly. We defer a discussion of the unscrambling of the results until the end of the following Section which describes a general algorithm for $n=2^k$.

2.3 The case $n = 2^k$

We write r and j in k-bit binary form

$$r = 2^{k-1} r_{k-1} + 2^{k-2} r_{k-2} + \ldots + r_0$$

$$j = 2^{k-1} j_{k-1} + 2^{k-2} j_{k-2} + \ldots + j_0$$

so that

$$X(j_{k-1}, j_{k-2}, \ldots, j_0) = \tag{2.10}$$

$$\sum_{r_0} \sum_{r_1} \ldots \sum_{r_{k-1}} A(r_{k-1}, r_{k-2}, \ldots, r_0) W_n^{j(2^{k-1} r_{k-1} + 2^{k-2} r_{k-2} + \ldots + r_0)},$$

where $W_n = \exp(2\pi i/n)$ and the summations are over $r_\nu = 0, 1$. Now it is easy to show that

$$W_n^{j 2^{k-1} r_{k-1}} = W_n^{j_0 2^{k-1} r_{k-1}}$$

and hence the innermost sum (over r_{k-1}) depends only on $j_0, r_{k-2}, r_{k-3}, \ldots, r_0$ and may be written

$$A_1(j_0, r_{k-2}, r_{k-3}, \ldots, r_0) = \sum_{r_{k-1}} A(r_{k-1}, r_{k-2}, \ldots, r_0) W_n^{j_0 2^{k-1} r_{k-1}}.$$

$$\tag{2.11}$$

Proceeding to the next innermost sum over r_{k-2} and so on and using

$$W_n^{j 2^{k-\kappa} r_{k-\kappa}} = W_n^{(j_{\kappa-1} 2^{\kappa-1} + \ldots + j_0) 2^{k-\kappa} r_{k-\kappa}}$$

we obtain the general recurrence relation

$$A_\kappa(j_0, j_1, \ldots, j_{k-1}, r_{k-\kappa-1}, \ldots, r_0) = \tag{2.12}$$

$$\sum_{r_{k-\kappa}} A_{\kappa-1}(j_0, j_1, \ldots, j_{k-2}, r_{k-\kappa}, \ldots, r_0) W_n^{(j_{k-1}2^{\kappa-1} + \ldots + j_0)2^{k-\kappa} r_{k-\kappa}}$$

for $\kappa = 1, 2, \ldots, k$, where we define $A_0(r) = A(r)$, $r = 0, 1, \ldots, n-1$. The last array calculated gives the desired results in a scrambled order, so that

$$X(j_{k-1}, j_{k-2}, \ldots, j_0) = A_k(j_0, j_1, \ldots, j_{k-1}). \tag{2.13}$$

The so called 'dual node' computation described for n=4 and characterised by (2.9) can be generalised to the present case. For n=4 the relevant pairs of array elements $[A_1(0), A_1(2)]$ and $[A_1(1), A_1(3)]$ are such that the difference in subscript for each pair is 2 (=n/2). Similarly at the second stage the appropriate pairs are $[A_2(0), A_2(1)]$ and $[A_2(2), A_2(3)]$, each with relevant subscript difference of 1 (=n/2^2). A similar analysis of the case n=8=2^3 leads one to expect that for the general case the appropriate pair of array elements at the κ-th stage is $[A_{\kappa-1}(r), A_{\kappa-1}(r+n/2^\kappa)]$ and that if the weighting factor in the equation for $A_\kappa(r)$ is W_n^p, that for $A_\kappa(r+n/2^\kappa)$ is $W_n^{p+n/2} = -W_n^p$. Hence we may write

$$\left. \begin{aligned} A_\kappa(r) &= A_{\kappa-1}(r) + W_n^p A_{\kappa-1}(r+n/2^\kappa) \\ A_\kappa(r+n/2^\kappa) &= A_{\kappa-1}(r) - W_n^p A_{\kappa-1}(r+n/2^\kappa). \end{aligned} \right\} \tag{2.14}$$

Thus if we work from the top down in the array $A_\kappa(r)$ (that is starting with r=0) we compute (2.14) for the first $n/2^\kappa$ elements, skip the next $n/2^\kappa$ and so on.

The computation of p may be carried out as follows. From (2.12) we see that at the κ-th stage

$$p = 2^{k-1} j_{\kappa-1} + 2^{k-2} j_{\kappa-2} + \ldots + 2^{k-\kappa} j_0, \qquad (2.15)$$

and the index r is given, in k-bit form, as

$$r = (j_0, j_1, \ldots, j_{\kappa-1}, r_{k-\kappa-1}, \ldots, r_0). \qquad (2.16)$$

Sliding this binary number $k-\kappa$ bits to the right and replacing all $k-\kappa$ bit positions on the left with zeros (integer division by $2^{k-\kappa}$) we obtain

$$\frac{r}{2^{k-\kappa}} = (0, 0, \ldots, 0, j_0, j_1, \ldots, j_{\kappa-1}) \ .$$

Reversing the order of the bits we find

$$(j_{\kappa-1}, j_{\kappa-2}, \ldots, j_0, 0, 0, \ldots, 0) = 2^{k-1} j_{\kappa-1} + 2^{k-2} j_{\kappa-2} + \ldots + 2^{k-\kappa} j_0 = p.$$
$$(2.17)$$

The unscrambling of the results is carried out essentially by reversing the bits of the indices of the final computed array given by (2.13). If we work top down in the array $A_k(r)$ we compute $\bar{r}$, the integer obtained by bit-reversing r, and interchange $A_k(r)$ and $A_k(\bar{r})$. In order to avoid the possibility of more than one interchange being carried out on the same pair of elements of A_k, at each stage we examine $\bar{r}$ to check if $\bar{r} < r$. If this is the case the interchange is not performed.

We conclude this Section by noting that this algorithm requires nk/2 complex multiplications and nk complex additions whereas direct evaluation of (2.1) requires n^2 complex multiplications and $n(n-1)$ complex additions. The ratio of complex multiplications required by the FFT algorithm to those required by a direct calculation is thus $2n/k$ which has a value approximately 21 for n=64 and a value greater than 200 for n=1024.

2.4   Discussion

Many variations on the basic FFT algorithm just described are to be found in the literature.   A good general reference is the book by Brigham (1974) in which there are Fortran and Algol programs (pp. 164-5) which implement the method of Section 2.3 (using real arrays) and many references to the FFT programs of various authors.   The bit manipulations required by the algorithm are best carried out using a low-level language, although implementations using purely high-level languages are suitable for many practical purposes and Brigham's Fortran program is reproduced in Appendix 2 .   Most large computer installations have associated with them scientific software packages which frequently include FFT programs and some microcomputers have available special hardware which allows FFT computation.

Some of the variations on, and improvements to, the algorithm given in the previous Section are now briefly considered.   We recall that the algorithm requires the input data in the natural order and produces the results in scrambled order and also that the powers of W_n are required in bit-reversed order.   If it is

arranged that the input data is in scrambled order then the output
is in the natural order and the powers of W_n occur in the natural
order.   A variation on the algorithm of Section 2.3 was given by
Gentleman and Sande (1966) in which the components of j rather than
r are separated.   Given the data in the natural order the results
are in scrambled order and the powers of W_n occur in natural order.
It is possible, of course, to obtain the results in natural order
using this algorithm if the input data is supplied in bit-reversed
order (the powers of W_n then occur in bit-reversed order).   FFT
algorithms which allow both input data and results to be in natural
order have been developed [see for example Cochran et al (1967)]
but twice the storage of the algorithms just described is required.
This, of course, may or may not be a practical disadvantage
depending on the amount of storage available.

Algorithms using a base other than 2 have been studied.   For
example, a base 4 algorithm can be formulated in a manner similar
to the development of the preceding Sections.   In particular it
is instructive for the reader to write down the base 4 algorithm
for $n=4^2$, and note the reduction of 25 per cent. in the number of
multiplications as compared with the corresponding base 2
algorithm.   A general mixed radix '4+2' algorithm has been
developed which computes as many steps as possible with the base 4
algorithm and the final step with a base 2 algorithm.   This
composite algorithm is popular since the restrictions on the choice
of n are the same as for a base 2 algorithm but the overall
efficiency is greater and Temperton (1983a,b,c) gives a detailed

account of some mixed-radix FFT algorithms. Dobes (1982) discusses
FFT algorithms with recursively generated trigonometric functions.

Operations counts have been given by Bergland (1968) for n any
power of 2 and show that in general the higher the base of the
algorithm the greater the computational efficiency which is
achieved. However as the base increases in value the complexity
of the algorithm increases. The operation counts given by
Bergland assume that each algorithm is written to perform as
efficiently as possible and, in particular, includes 'twiddling' or
'referencing' operations to achieve this efficiency. The
essential idea of 'twiddle' factors is illustrated in the following
discussion. From Section 2.2 we recall the algorithm for $n=2^2$
given by equation (2.3) which we re-state below.

$$X(j_1,j_0) = \sum_{r_0=0}^{1} \left[\sum_{r_1=0}^{1} A(r_1,r_0)W_4^{2j_0r_1} \right] W_4^{(2j_1+j_0)r_0}. \tag{2.18}$$

In the above equation and in relations (2.4) and (2.5) the term
$W_4^{j_0r_0}$ was grouped arbitrarily with the outer sum and an alternative
approach is to group this term with the inner sum so that (2.18)
may be split in the form

$$A_1(j_0,r_0) = \left[\sum_{r_1=0}^{1} A(r_1,r_0)W_4^{2j_0r_1} \right] W_4^{j_0r_0}, \tag{2.19}$$

$$A_2(j_0,r_0) = \sum_{r_0=0}^{1} A_1(j_0,r_0)W_4^{2j_1r_0}, \tag{2.20}$$

and

$$X(j_1,j_0) = A_2(j_0,j_1). \tag{2.21}$$

We note that

$$W_4^{2j_0r_1} = e^{\pi i j_0 r_1}$$

so that this term only takes the values ± 1.   Consequently the sum
in brackets in (2.19) may be evaluated without multiplications and
the results 'twiddled' by the factor $W_4^{j_0r_0}$.   Each twiddling
operation requires one complex multiplication, except of course
when the factor is W_4^0.   Since equation (2.20) may similarly be
evaluated without multiplications it is clear that the overall
number of multiplications is reduced compared with the algorithm of
Section 2.2.   This reduction has been brought about essentially by
exploiting the symmetries of the sine and cosine functions and
these ideas can be applied to more general algorithms with
consequent improvements in efficiency.   Algorithms for n having
arbitrary factors have been developed and Korn and Lambrotte
(1979), Fornberg (1981) and Temperton (1983a,b,c) have implemented
FFT programs on vector computers.

    In Sections 2.5 to 2.11 we assume that there is available a
routine for computing the Fourier series (IDFT)

$$X(j) = \sum_{r=0}^{n-1} A(r)W_n^{jr} \quad (j=0,1,\ldots,n-1) \qquad (2.22)$$

which is based on an FFT algorithm.   As noted in Chapter 1, any
algorithm which computes (2.22) can be used to compute the discrete
Fourier transform (DFT)

$$A(r) = \frac{1}{n} \sum_{j=0}^{n-1} X(j)W_n^{-rj} \quad (r=0,1,\ldots,n-1) \qquad (2.23)$$

by replacing $A(r)$ by $X^*(j)/n$ in (2.22) and taking the complex
conjugate of the result. Some properties of the sequence $X(j)$
and $A(r)$ are next considered in order to facilitate developments in
subsequent Sections.

2.5 Some useful definitions and properties

A basic property of the sequences $X(j)$ and $A(r)$ defined by
(2.22) and (2.23) is their periodicity. From (2.22) we see that,
for a given value of j and any integer ν

$$X(j + n\nu) = \sum_{r=0}^{n-1} A(r)W_n^{jr}W_n^{n\nu r}$$

$$= X(j), \quad \nu=0,\pm1,\pm2,\ldots \qquad (2.24)$$

since $W_n^{n\nu r} = 1$. Similarly from (2.23), for a given value of r

$$A(r+n\nu) = A(r), \quad \nu=0,\pm1,\pm2,\ldots \qquad (2.25)$$

Thus both sequences (2.22) and (2.23) are each periodic with a
period of n sequence elements.

Some special types of sequence will be encountered in the
following Sections. In particular we note that a sequence is
called even if

$$X(j) = X(n-j) \quad (j=0,1,\ldots,n-1) \qquad (2.26)$$

and odd if

$$X(j) = -X(n-j) \quad (j=0,1,\ldots,n-1). \qquad (2.27)$$

A sequence is respectively conjugate even or conjugate odd if

$$X(j) = X^*(n-j) \quad (j=0,1,\ldots,n-1) \qquad (2.28)$$

or

$$X(j) = -X^*(n-j) \quad (j=0,1,\ldots,n-1). \qquad (2.29)$$

It is easy to show that if $X(j)$ is conjugate even then all the $A(r)$ are real and if $X(j)$ is conjugate odd, then the $A(r)$ are purely imaginary. These results are also true if one interchanges the roles of $A(r)$ and $X(j)$ as is clear from the relations (2.22) and (2.23). As an illustration we show that if $A(r)$ is conjugate even then $X(j)$ is real, a result which will be of particular utility later when working with real sequences.

We note first that if $A(r)$ is conjugate even $A(n/2)=A^*(n/2)$ so that $A(n/2)$ must be real and that $A(0)$ also must be real, since the sequence $A(r)$ is conjugate even and periodic. (These results are obvious from (2.23) if $X(j)$ is real). Furthermore, for any integer μ

$$W_n^{j(n-\mu)} = e^{-2\mu\pi i j/n} = W_n^{-j\mu}, \qquad (2.30)$$

and hence it is easy to show that all imaginary parts in the summation (2.22) cancel and we may write

$$X(j) = \sum_{r=0}^{n/2} \left[a(r)\cos(2\pi jr/n) + b(r)\sin(2\pi jr/n) \right] \qquad (2.31)$$

where

$$a(0) = A(0), \quad a(n/2) = A(n/2) \qquad (2.32)$$

and

$$a(r) = 2 \, \text{Re}\big[A(r)\big], \; b(r) = -2 \, \text{Im}\big[A(r)\big]. \quad (r=1,2,\ldots,n/2-1) \quad (2.33)$$

Many other properties can be deduced for various specialised $X(j)$ and $A(r)$. For example $X(j)$ is real and even if and only if $A(r)$ is real and even. Similarly $X(j)$ is real and odd if and only if $A(r)$ is imaginary and odd. The last property we state here is that the IDFT and DFT are linear. This is apparent from the definitions (2.22) and (2.23) since if

$$A(r) = \alpha A_1(r) + \beta A_2(r), \quad (r=0,1,\ldots,n-1) \quad (2.34)$$

where α and β are constants and

$$X_1(j) = \sum_{r=0}^{n-1} A_1(r) W_n^{jr} \quad\left.\right] \quad (2.35)$$

$$X_2(j) = \sum_{r=0}^{n-1} A_2(r) W_n^{jr}$$

then

$$X(j) = \alpha X_1(j) + \beta X_2(j). \quad (2.36)$$

2.6 The Fourier transform of two real data sequences using a
   ## complex DFT routine

Suppose that $X_1(j)$ and $X_2(j)$ are two n-point sequences such that

$$A_1(r) = \frac{1}{n} \sum_{j=0}^{n-1} X_1(j) W_n^{-rj}, \quad A_2(r) = \frac{1}{n} \sum_{j=0}^{n-1} X_2(j) W_n^{-rj} \quad (2.37)$$

for $r=0,1,\ldots,n-1$. We let

$$X(j) = X_1(j) + iX_2(j) \quad (j=0,1,\ldots,n-1) \quad (2.38)$$

so that

$$A(r) = A_1(r) + iA_2(r), \qquad (2.39)$$

where

$$A(r) = \frac{1}{n} \sum_{j=0}^{n-1} X(j)W_n^{-rj}, \qquad (2.40)$$

for $r=0,1,\ldots,n-1$ and since $X_1(j)$ and $X_2(j)$ are real we find that

$$\left.\begin{array}{l} A_1(r) = A_1^*(n-r) \\[2mm] A_2(r) = A_2^*(n-r). \end{array}\right\} \qquad (2.41)$$

Taking the complex conjugate of (2.39), replacing r by n-r and using (2.41) we obtain

$$A^*(n-r) = A_1(r) - iA_2(r). \qquad (2.42)$$

Equations (2.39) and (2.42) may thus be solved for $A_1(r)$ and $A_2(r)$ giving

$$\left.\begin{array}{l} A_1(r) = \frac{1}{2}\left[A^*(n-r) + A(r)\right] \\[2mm] A_2(r) = \frac{i}{2}\left[A^*(n-r) - A(r)\right]. \end{array}\right\} \qquad (2.43)$$

Hence the overall computational procedure is that given $X_1(j)$ and $X_2(j)$ we form $X(j)$ via (2.38), compute $A(r)$ [the complex DFT of $X(j)$] and compute $A_1(r)$ and $A_2(r)$ from (2.43) for $r = 0,1,\ldots,n/2$.

Values for r>n/2 need not be computed since they are immediately available from (2.41) and it should be noted that $A^*(n)$, which is required in the evaluation of (2.43) for r=0, may as a result of (2.25) be replaced by $A^*(0)$ $[=A(0)]$.

2.7 The Fourier transform of 2n (complex) points from two separate n-point transforms (the 'doubling' algorithm)

Suppose we have 2n complex data values Y(j), j=0,1,...,2n-1 such that

$$C(r) = \frac{1}{2n} \sum_{j=0}^{2n-1} Y(j)W_{2n}^{-rj}, \quad (r=0,1,\ldots,2n-1). \qquad (2.44)$$

We define

$$\left.\begin{array}{l} X_1(j) = Y(2j) \\[2mm] X_2(j) = Y(2j+1) \end{array}\right] \quad (j=0,1,\ldots,n-1), \qquad (2.45)$$

that is, we separate the even and odd subscripted elements of Y(j) and denote by respectively $A_1(r)$ and $A_2(r)$ (r=0,1,...,n-1) the DFTs of $X_1(j)$ and $X_2(j)$. We write (2.44) in the form

$$C(r) = \frac{1}{2n}\left[\sum_{j=0}^{n-1} Y(2j)W_{2n}^{-2rj} + \sum_{j=0}^{n-1} Y(2j+1)W_{2n}^{-2rj-r}\right] \qquad (2.46)$$

and note that since $W_{2n}^2 = W_n$ equation (2.46) may be written as

$$C(r) = \frac{1}{2}\left[A_1(r) + A_2(r)W_{2n}^{-r}\right], \tag{2.47}$$

which is defined for r=0,1,...,n-1. To compute C(r) for

r=n,n+1,...,2n-1 we replace r by n+r in (2.47), make use of the

periodicity of $A_1(r)$ and $A_2(r)$ and the fact that $W_{2n}^{-n} = -1$ to obtain

$$C(n+r) = \frac{1}{2}\left[A_1(r) - A_2(r)W_{2n}^{-r}\right]. \tag{2.48}$$

Thus we compute the two n-point DFTs $A_1(r)$ and $A_2(r)$ of the

sequences defined by (2.45) and use (2.47) and (2.48) for

r=0,1,...,n-1 to compute the 2n-point DFT of Y(j).

It is worth noting here that this algorithm, when suitably

iterated upon to produce successive doublings up to the total

number of points required, is the FFT algorithm with base 2. For

example, if we were to use the algorithm to compute the DFT of an

eight-point sequence $\left[Y(0),Y(1),...,Y(7)\right]$, we would first form the

two four-point sequences $\left[Y(0),Y(2),Y(4),Y(6)\right]$ and

$\left[Y(1),Y(3),Y(5),Y(7)\right]$, as defined by (2.45). Each of these

four-point sequences may be transformed as two, two-point

sequences. Thus from the first four-point sequence we form the

two-point sequences $\left[Y(0),Y(4)\right]$ and $\left[Y(2),Y(6)\right]$ and from the second

four-point sequence the sequences $\left[Y(1),Y(5)\right]$ and $\left[Y(3),Y(7)\right]$.

Thus at first stage of computation four two-point transforms are

performed and at the next stage these results are used to form two

four-point transforms. The final stage uses these two results to

form the required eight-point transform. A Fortran program based
on this method is given, for example, in Gonzalez and Wintz
(1977,p87). It is easy to verify that the re-ordered data
sequence $[Y(0),Y(4),Y(2),Y(6),Y(1),Y(5),Y(3),Y(7)]$ is obtained from
the original sequence simply by reversing the bits of the indices,
as described in Section 2.3.

2.8   The Fourier transform of 2n real data points

Essentially we carry out the 2n-point transform of the real
data $Y(j)$, $j=0,1,\ldots,2n-1$ by means of one DFT of n complex points
using the results of Sections 2.6 and 2.7. Additionally we make
use of the fact that the Fourier harmonics $C(r)$ are conjugate even
$[C(r)=C^*(2n-r)]$, where $C(r)$ is given by (2.44). The formula for
$Y(j)$ in terms of $C(r)$ may thus be written in a form analogous to
(2.31) so that

$$Y(j) = \frac{1}{2} a(0) + \sum_{r=1}^{n-1} \left[a(r)\cos(\pi jr/n) + b(r)\sin(\pi jr/n) \right] + \frac{1}{2} (-1)^j a(n) \tag{2.49}$$

where

$$\left. \begin{array}{l} a(r) = 2\mathrm{Re}\left[C(r) \right], \quad (r=0,1,\ldots,n) \\[2mm] b(r) = -2\mathrm{Im}\left[C(r) \right], \quad (r=1,2,\ldots,n-1) \end{array} \right] \tag{2.50}$$

This type of transform arises from problems with periodic boundary
conditions, as indicated in Appendix 1 (Section A1.4), and here we
describe an efficient calculation procedure.

We form the n-point sequences $X_1(j)$ and $X_2(j)$ given by (2.45) (now real) and the complex sequence

$$X(j) = X_1(j) + iX_2(j), \quad (j=0,1,\ldots,n-1) \qquad (2.51)$$

in the same manner as equation (2.38) of Section 2.6. It is

usually the case (e.g. in Fortran) that complex arrays are stored

so that real and imaginary parts are in alternate locations and

hence a real 2n-point array can be naturally put into an n-point

complex array $X(j)$ defined by (2.51). Alternatively, separate real

arrays may be used for the real and imaginary parts of $X(j)$, as in

Brigham's program, Appendix 2 .

The transforms of respectively $X_1(j)$ and $X_2(j)$ are conjugate

even and are given by (2.43) for $r=0,1,\ldots,n/2$. Relations (2.47)

and (2.48) may then be used to compute $C(r)$ and $C(n+r)$

respectively, noting that each relation needs to be used only for

$r=0,1,\ldots,n/2$ in view of the fact that $C(r)=C^*(2n-r)$, so that

$C(n/2+\nu)=C^*(3n/2-\nu)$, $(\nu=1,2,\ldots,n/2-1)$. Thus the upper half

$(n<r\leq 2n-1)$ of the array $C(r)$ is redundant and need not be computed

or stored. We note also that since $C(0)$ and $C(n)$ are real, our

data sequence of 2n real values has been transformed to a sequence

of harmonics which contains only 2n independent real numbers.

2.9   Calculation of Fourier series for real data

Here we are concerned with evaluating (2.49) given the Fourier

harmonics $C(r)$ which are conjugate even. Essentially we use the

methods of Sections 2.7 and 2.8 in reverse. From (2.47) and

(2.48) we obtain

$$A_1(r) = C(r) + C(n+r)$$

$$A_2(r) = \left[C(r) - C(n+r) \right] W_{2n}^r$$

(2.52)

and from (2.43) we find

$$A(r) = A_1(r) + iA_2(r)$$

(2.53)

and

$$A^*(n-r) = A_1(r) - iA_2(r).$$

(2.54)

Thus we calculate $A_1(r)$ and $A_2(r)$ for $r=0,1,\ldots,n/2$ from (2.52) [remembering that only values of $C(r)$ for $r=0,1,\ldots,n$ need be supplied since $C(n+r)=C^*(n-r)$] and generate the complete set of complex values $A(r)$, $r=0,1,\ldots,n-1$ by using (2.53) and (2.54) for $r=0,1,\ldots,n/2$. The IDFT of the sequence $A(r)$, $r=0,1,\ldots,n-1$ is the n-point complex sequence $X(j)$ whose real and imaginary parts are the elements of the 2n-point real sequence $Y(j)$ defined by (2.45).

2.10 Calculation of cosine series for real data

It is easy to show that if $Y(j)$, $j=0,1,\ldots,2n-1$ is real and even so that $Y(j)=Y(2n-j)$, then its IDFT, given by

$$Y(j) = \sum_{r=0}^{2n-1} C(r)W_{2n}^{jr}$$

is expressible as a cosine series in the form

$$Y(j) = \frac{1}{2} a(0) + \sum_{r=1}^{n-1} a(r)\cos(\pi jr/n) + \frac{1}{2} (-1)^j a(n), \qquad (2.55)$$

where

$$a(r) = 2C(r) \qquad (2.56)$$

and we recall from Section 2.5 that the harmonics $C(r)$, $r=0,1,\ldots,2n-1$ must also be real and even. Thus the DFT of $Y(j)$ may be written as the cosine series

$$a(r) = \frac{2}{n} \left[\frac{1}{2} Y(0) + \sum_{j=1}^{n-1} Y(j)\cos(\pi rj/n) + \frac{1}{2} (-1)^r Y(n) \right]. \qquad (2.57)$$

As indicated in Chapter 1 and Appendix 1 (Section A1.3), the above form of series arises in problems with Neumann boundary conditions and here we describe an efficient method for calculating the harmonics $a(r)$, $r=0,1,\ldots,n$.

We define the complex sequence

$$X(j) = Y(2j) + i\left[Y(2j+1) - Y(2j-1) \right], \qquad (2.58)$$

for $j=0,1,\ldots,n-1$ and it is easy to verify that this is a conjugate even sequence whose transform $A(r)$ must therefore be real. In Section 2.9 we described an efficient method for calculating the Fourier series for a 2n-point real sequence given the conjugate even Fourier harmonics $C(r)$ of which only values for $r=0,1,\ldots,n$ need be supplied. It is thus clear that by supplying the values $X^*(j)/n$, $j=0,1,\ldots,n/2$ as the given values of $C(r)$ of Section 2.9

44

(which of course also implies that n is replaced by n/2 throughout
that Section)   the output will be real values A(r),
r=0,1,...,n-1.

The method of Section 2.6 can be used to obtain the transforms
of the real and imaginary parts of X(j).   We define

$$A_1(r) = \frac{1}{n} \sum_{j=0}^{n-1} Y(2j)W_n^{-rj} \qquad (2.59)$$

and

$$A_3(r) = \frac{1}{n} \sum_{j=0}^{n-1} \left[Y(2j+1) - Y(2j-1) \right]W_n^{-rj}, \qquad (2.60)$$

so that equations (2.43) may be written (remembering that A(r) is
real) as

$$A_1(r) = \frac{1}{2} \left[A(n-r) + A(r) \right] \qquad (2.61)$$

and

$$A_3(r) = \frac{i}{2} \left[A(n-r) - A(r) \right]. \qquad (2.62)$$

We introduce the further definition

$$A_2(r) = \frac{1}{n} \sum_{j=0}^{n-1} Y(2j+1)W_n^{-rj} \qquad (2.63)$$

and note that since the Fourier harmonics associated with Y(2j-1)
are $A_2(r)W_n^{-r}$ we may derive the relation

$$A_2(r) = A_3(r)/(1 - W_n^{-r}). \qquad (2.64)$$

Thus the doubling algorithm of Section 2.7 may be used to calculate
the required C(r) values by substituting (2.61) and (2.64) into
(2.47) giving

$$a(r) = 2C(r) = \frac{1}{2}\left[A(n-r) + A(r)\right] + \frac{i}{2}\left[A(n-r) - A(r)\right]\frac{W_{2n}^{-r}}{1-W_n^{-r}}$$

$$= \frac{1}{2}\left[\left[A(n-r) + A(r)\right] + \left[A(n-r) - A(r)\right]/\left[2\sin(\pi r/n)\right]\right], \qquad (2.65)$$

which may be used for $r=1,2,\ldots,n-1$. For $C(0)$ and $C(n)$ we must make a special calculation. Equations (2.47) and (2.48) with $r=0$ are

$$2C(0) = A_1(0) + A_2(0) \qquad (2.66)$$

and

$$2C(n) = A_1(0) - A_2(0), \qquad (2.67)$$

where $A_1(0)$ is immediately available from (2.61) and $A_2(0)$ may be obtained directly from the definition

$$A_2(0) = \frac{1}{n}\sum_{j=0}^{n-1} Y(2j+1). \qquad (2.68)$$

This procedure effectively reduces by a factor of four the computation and storage requirements compared with the straightforward approach since, in particular, the complex array to be transformed contains only $n/2$ elements instead of the $2n$ which would be required if the full array $Y(j)$ were supplied as input to a complex DFT routine.

2.11 Calculation of sine series for real data

If $Y(j)$, $j=0,1,\ldots,2n-1$, is real and odd so that $Y(j)=-Y(2n-j)$, it is a straightforward matter to show that its IDFT

$$Y(j) = \sum_{r=0}^{2n-1} C(r)W_{2n}^{jr}$$

may be written as the sine series

$$Y(j) = \sum_{r=1}^{n-1} b(r)\sin(\pi jr/n), \qquad (2.69)$$

where

$$b(r) = 2iC(r). \qquad (2.70)$$

We note that all the values of $b(r)$, $r=1,2,\ldots,n-1$ are real (the $C(r)$ values $r=0,1,\ldots,2n-1$ are imaginary and odd). It is clear that $Y(0)=Y(n)=0$ and that

$$b(r) = \frac{2}{n} \sum_{j=1}^{n-1} Y(j)\sin(\pi rj/n), \quad (r=1,2,\ldots,n-1) \qquad (2.71)$$

and

$$b(0) = b(n) = 0. \qquad (2.72)$$

As indicated in Chapter 1 and Appendix 1 (Section A1.2), this form of series arises in problems with Dirichlet boundary conditions and we describe an efficient method for evaluating the harmonics (2.71) which is very similar to the method described in the preceding Section for the cosine series.

We begin by defining for $j=0,1,\ldots,n-1$ the conjugate even sequence

$$X(j) = \left[Y(2j+1) - Y(2j-1) \right] + iY(2j), \qquad (2.73)$$

By supplying $X^*(j)/n$ for $j=0,1,\ldots,n/2$ as the values of $C(r)$ of Section 2.9 we obtain the n real values $A(r)$, $r=0,1,\ldots,n-1$, the transform of $X(j)$. If we define $A_1(r)$, $A_2(r)$ and $A_3(r)$ by respectively (2.59), (2.63) and (2.60) then the equations corresponding to (2.61) and (2.62) become

$$A_3(r) = \frac{1}{2}\left[A(n-r) + A(r)\right]$$

(2.74)

and

$$A_1(r) = \frac{i}{2}\left[A(n-r) - A(r)\right]$$

respectively. Thus using (2.47), (2.64) and (2.74) we may deduce that

$$b(r) = 2iC(r) = \frac{1}{2}\left[\left[A(r) - A(n-r)\right] + \left[A(r) + A(n-r)\right]/\left[2\sin(\pi r/n)\right]\right]$$

(2.75)

for $r=1,2,\ldots,n-1$. It is clear that a computational efficiency similar to that for the cosine series evaluation of the preceding Section is achieved by the use of the above procedure.

An alternative calculation procedure for the sine transform was given by Temperton (1980) in which the division by $\sin(\pi r/n)$ in (2.75) is avoided. The essential point to note is that, apart from a scaling factor, the sine transform is its own inverse, as indicated by equations (2.71) and (2.69). Thus replacing $b(r)$ by $Y(j)$ and $A(r)$ by $Z(j)$ we can invert equation (2.75) to give

$$Z(j) = \left[Y(j) + Y(n-j)\right]\sin(\pi j/n) + \frac{1}{2}\left[Y(j) - Y(n-j)\right],$$

(2.76)

for $j=1,2,\ldots,n-1$, and it is clear that

$$Z(0) = 0.$$

(2.77)

The real n-point sequence defined by (2.76) and (2.77) may be input to the algorithm of Section 2.8, the output of which is thus $n/2+1$ elements of an $n+1$ point conjugate even sequence $\bar{C}(r)$. The values

of $b(r)$ may be deduced from the form of the right-hand side of
(2.73), since essentially $X(j)$ in (2.73) is to be replaced by
$\bar{C}^*(r)$. Hence, taking account of the scaling factor, we find

$$b(2r) = -2\text{Im}\left[\bar{C}(r)\right] \qquad (2.78)$$

and

$$b(2r+1) = b(2r-1) + 2\text{Re}\left[\bar{C}(r)\right], \qquad (2.79)$$

for $r = 1,2,\ldots,n/2-1$ where

$$b(1) = \bar{C}(0). \qquad (2.80)$$

Temperton (1980) examined the accuracy of solutions of
Poisson's equation with Dirichlet boundary conditions using FFT
methods with both forms of sine transform. For a given 'true'
numerical solution he determined the corresponding right-hand side
of the discrete Poisson equation and used this as numerical input
to his program PSOLVE. By comparing the resulting computed
solution with the 'true' solution for a number of cases he found
that the mean maximum absolute error was roughly proportional to n^2
if relation (2.75) were employed, whereas using (2.78) - (2.80) the
error was found to be roughly proportional to n. For n=128 the
error was reduced by approximately a factor of 10 using truncated
floating-point arithmetic on a CYBER-175.

Both forms of sine transform give rise to errors which are
effectively negligible for most practical purposes but since the
work involved in using either algorithm is essentially the same, it
is clearly worthwhile to use the modified version, especially for
large n. It seems probable that similar remarks should apply to
the cosine transform in view of the form of equation (2.65).

CHAPTER 3
FFT Solution of Partial Differential Equations

3.1 Introduction

This Chapter is primarily concerned with the application of FFT methods to various types of linear partial differential equations and we begin by discussing some aspects of linear algebra which are relevant to the classes of problems considered in Section 3.3 and subsequent Sections. We consider a system of linear algebraic equations

$$B\underset{\sim}{\psi} = \underset{\sim}{g}, \tag{3.1}$$

in which the matrix B is written in partitioned form as

$$B = \begin{bmatrix} B_{11} & B_{12} & \cdots & B_{1\sigma} \\ B_{21} & B_{22} & \cdots & B_{2\sigma} \\ \cdot & \cdot & \cdots & \cdot \\ B_{\sigma 1} & B_{\sigma 2} & \cdots & B_{\sigma\sigma} \end{bmatrix}, \tag{3.2}$$

where each of the σ^2 blocks $B_{\alpha\beta}$, $\alpha,\beta = 1,2,\ldots,\sigma$ is a square matrix of order ρ and the vectors $\underset{\sim}{\psi}$ and $\underset{\sim}{g}$ are each partitioned into σ vectors each of length ρ in the form

$$\underset{\sim}{\psi} = \begin{bmatrix} \psi_1, \psi_2, \ldots, \psi_\sigma \end{bmatrix}^T \quad \text{and} \quad \underset{\sim}{g} = \begin{bmatrix} g_1, g_2, \ldots, g_\sigma \end{bmatrix}^T . \tag{3.3}$$

For the Dirichlet problem formulated in Chapter 1 it is clear

49

that the block tridiagonal structure of equations (1.7) is a
special case of (3.1)-(3.3). More generally, for a two-dimensional
elliptic partial differential which is to be solved over a
rectangular region at each of $\rho\sigma$ grid points (arranged in σ rows
each containing ρ nodes) it is clear that the complete set of
finite difference equations (3.1) may be naturally partitioned into
the form of (3.2) and (3.3). The detailed structure of B of course
depends not only on the chosen form of approximation for the
partial differential equation, but also on the boundary
conditions.

For parabolic and hyperbolic problems in two independent
variables, one of which is usually time, t, the totality of
difference equations which approximate the differential problem up
to some finite time, t_f, also may be expressed in the form (3.1).
The structure of (3.2) is then lower triangular and the number of
null blocks in the lower triangle depends on the number of time
levels involved in the particular difference scheme under
consideration. For such problems the vectors $\underset{\sim}{\phi}_1$, $\underset{\sim}{\phi}_2$, ..., $\underset{\sim}{\phi}_f$ are
usually generated by a step-by-step procedure and some details are
given in later examples.

In Chapter 1 it was noted that part of the efficiency of the
overall method is achieved through the fact that the eigenvectors
of the relevant matrix A [equation (1.8)] are mutually orthogonal.
For the more general system (3.1), it is clear that in order to
decouple our system of $\rho\sigma$ equations into ρ independent sets of
equations each of order σ, we require all the submatrices $B_{\alpha\beta}$ to
have the same set of mutually orthogonal eigenvectors. It can be

proved that this is the case if all these submatrices commute and are, in addition, real symmetric matrices (see, for example, Varga p220). Hence the decoupling of our system can be guaranteed for a matrix B having the block structure (3.2) and the above properties. For time-dependent problems which are approximated by implicit difference formulae, the above requirements make the computational procedure explicit, as illustrated by the examples of Sections 3.3.3 and 3.3.4.

In practice the decomposition method shows to its best advantage when the relevant eigenvectors have components which are sines and/or cosines so that the techniques described in the previous Chapter may be fully utilised. Thus we are led to consider a particular class of linear partial differential equations and boundary conditions whose discrete representations generate suitable coefficient matrices B. It is worth noting here that for some types of boundary conditions (eg Neumann conditions) the blocks are not symmetric but are simply related to symmetric matrices. In such cases the eigenvectors are not simply mutually orthogonal but are mutually orthogonal with respect to certain prescribed positive weights and some details are given in the following discussion and in Appendix 1. More general separable problems are briefly considered in Section 3.6.

3.2 Discrete representations; general considerations; boundary conditions.

We consider a two-dimensional second-order partial differential equation satisfied over a rectangular region for which a discrete representation at a grid point (j,s) may be written in a general nine-point form as

$$\sum_{\mu,\nu=-1}^{1} b_{\mu,\nu}(j,s)\, \phi_{j+\mu,s+\nu} = \Gamma_{j,s}, \tag{3.4}$$

and assume that the DFT is to be performed in the direction

corresponding to index j. For the components of the eigenvectors

of the relevant submatrices to consist of suitable sine or cosine

terms we require that the coefficients $b_{\mu,\nu}$ are independent of j

and that the coefficients $b_{\pm 1,\nu}$ are symmetric (or at least can be

made symmetric) with respect to index j, so that $b_{1,\nu}(s) =$

$b_{-1,\nu}(s)$. Thus with these conditions it follows that a typical

block of (3.2) may be written as

$$\begin{bmatrix} b_{0,\nu}(s) & b_{1,\nu}(s) & & & \\ b_{1,\nu}(s) & b_{0,\nu}(s) & b_{1,\nu}(s) & & \\ & \cdot & \cdot & \cdot & \cdot \\ & & b_{1,\nu}(s) & b_{0,\nu}(s) & b_{1,\nu}(s) \\ & & & b_{1,\nu}(s) & b_{0,\nu}(s) \end{bmatrix}, \tag{3.5}$$

where here, for simplicity, we assume that ϕ is zero on the

boundary. Such a matrix is clearly a simple linear function of

the "basic" matrix

$$\begin{bmatrix} -2 & 1 & & & \\ 1 & -2 & 1 & & \\ & \cdot & \cdot & \cdot & \\ & & 1 & -2 & 1 \\ & & & 1 & -2 \end{bmatrix}, \tag{3.6}$$

which was encountered previously (Chapter 1) for the special case

of the discrete Poisson equation with Dirichlet boundary
conditions.

As already indicated, the precise structure of a block is
dependent on the boundary conditions and we now consider this point
in more detail. We assume that the region of interest is covered
by a regular grid defined by the indices $j = 0,1,\ldots,R+1$ and
$s = 0,1,2,\ldots,S+1$ so that the boundaries correspond to
$j = 0,R+1$ and $s = 0,S+1$, and consider the following forms of
discretised conditions on $j = 0,R+1$.

$$\phi_{0,s} = f_s \quad , \quad \phi_{R+1,s} = F_s \quad , \quad (3.7a)$$

$$\phi_{1,s} - \phi_{-1,s} = 2h_x g_s \quad , \quad \phi_{R+2,s} - \phi_{R,s} = 2h_x G_s, \quad (3.7b)$$

$$\phi_{0,s} = f_s \quad , \quad \phi_{R+2,s} - \phi_{R,s} = 2h_x G_s, \quad (3.7c)$$

$$\phi_{R+1+\eta,s} = \phi_{\eta,s} \quad (\eta = 0,\pm 1,\pm 2\ldots), \quad (3.7d)$$

where f_s, g_s, F_s, G_s are prescribed functions and h_x is the
(constant) grid length in the j-direction. The right-hand sides of
(3.7b,c) have been written in this form for notational convenience
in subsequent Sections. On the boundaries $s = 0,S+1$ any boundary
conditions which are appropriate to the type of equation being
solved may be chosen and this choice affects the value of σ in
(3.2).

For conditions (3.7a) it is implied that terms in the relevant
difference equations which contain these known values are taken to
the right-hand side. Thus problems with such conditions, with, in
general, non-zero boundary values are solved in exactly the same
way as a problem with $f_s = F_s \equiv 0$, but with a modified right-hand
side, as illustrated by equations (1.4). Furthermore the "basic"

matrix for such problems is (3.6) which is of order $\rho = R$.

For (3.7b) the "fictitious" values $\phi_{-1,s}$ and $\phi_{R+2,s}$ may be eliminated using (3.4) for $j = 0, R+1$ in the standard fashion and known terms taken to the right-hand side. It then follows that the matrix corresponding to (3.6) has the form

$$\begin{bmatrix} -2 & 2 & & & & \\ 1 & -2 & 1 & & & \\ & & \cdot & \cdot & \cdot & \\ & & & 1 & -2 & 1 \\ & & & & 2 & -2 \end{bmatrix}, \qquad (3.8)$$

which is of order $\rho = R + 2$ and the appropriate eigenvectors of (3.8) are derived in Appendix I [Section A1.3]. It is worth noting here that a condition of the form

$$\phi_{1,s} - \phi_{-1,s} = 2h_x \, g_s \phi_{0,s},$$

does not allow a simple closed-form solution for the eigenvectors. Conditions (3.7c) are a mixture of (3.7a,b) and details for this case, for which $\rho = R + 1$, are given in Section A1.5.

The periodic conditions (3.7d) give rise to a "basic" matrix of the cyclic form

$$\begin{bmatrix} -2 & 1 & & & & 1 \\ 1 & -2 & 1 & & & \\ & & \cdot & \cdot & \cdot & \\ & & & 1 & -2 & 1 \\ 1 & & & & 1 & -2 \end{bmatrix}, \qquad (3.9)$$

of order $\rho = R + 1$, and Section A1.4 considers the relevant analysis for this case.

We close this Section by noting that for boundary conditions of the form (3.7a,b,c) we must choose R = n-1 where the value of n is suitable for FFT methods to be used (eg n = 2^k), whereas for (3.7d) R = 2n-1.

3.3 Two-dimensional second-order problems

3.3.1 General form soluble by FFT methods

The discussion of the previous Sections indicates that, for a function $\phi(x,y)$ which satisfies a linear equation of the form

$$\frac{\partial^2\phi}{\partial x^2} + a(y)\,\frac{\partial^2\phi}{\partial y^2} + b\,\frac{\partial\phi}{\partial x} + c(y)\,\frac{\partial\phi}{\partial y} + d(y)\,\phi = q(x,y), \quad (3.10)$$

FFT methods may be used (using the Fourier transform in the x-direction) to determine $\phi(x,y)$ provided that the solution domain is rectangular and the boundary conditions are of a suitable form. It should be clear to the reader that, apart from the inclusion of the term $b\partial\phi/\partial x$, with constant coefficient b, the particular separable form of (3.10) leads to difference equations with the desired properties if, for example, standard central difference approximations are used throughout for the derivatives. The details of how the term $b\partial\phi/\partial x$ may be accommodated will be illustrated in an example.

We suppose that the region of interest is finite in the x-direction, so that $0 \leqslant x \leqslant L_x$ but may be finite or infinite in

extent in the y-direction depending on the type of equation
represented by (3.10). Although for convenience the notation y is
used here for the second independent variable, this variable may
represent space or time. Thus the initial and/or boundary
conditions asociated with the y-direction will be given as
appropriate in later examples and, additionally, problems in other
coordinate systems may be written in the form (3.10) with suitable
interpretation of the variables x and y.

For our general equation (3.10) the allowable forms of
boundary conditions associated with the x-direction are

$$\phi(0,y) = f(y) \ , \quad \phi(L_x,y) = F(y), \qquad (3.11a)$$

$$\frac{\partial\phi(0,y)}{\partial x} = g(y) \ , \ \frac{\partial\phi(L_x,y)}{\partial x} = G(y), \qquad (3.11b)$$

$$\phi(0,y) = f(y) \quad , \ \frac{\partial\phi(L_x,y)}{\partial x} = G(y), \qquad (3.11c)$$

$$\phi(L_x,y) = \phi(0,y). \qquad (3.11d)$$

Equations (3.11a) are Dirichlet conditions and the Neumann
conditions (3.11b) are represented by (3.7b) if the usual
central-difference approximation with step-length h_x is used for
the derivative. Conditions (3.11c) with $G(y) \equiv 0$ are particularly
useful for symmetric Dirichlet problems and (3.11d) represent the
periodic conditions (3.7d).

The term $b \partial \phi / \partial x$ in (3.10), ($b \neq 0$) destroys the symmetry with respect to index j of the blocks of the matrix (3.2), as shown by the following illustrative example. For the equation

$$\nabla^2 \phi + b \frac{\partial \phi}{\partial x} = q(x,y) \tag{3.12}$$

a typical finite difference equation, assuming a square grid of side h, is

$$(1-2bh)\phi_{j-1,s} - 4\phi_{j,s} + (1+2bh)\phi_{j+1,s} + \phi_{j,s-1} + \phi_{j,s+1} = h^2 q_{j,s}, \tag{3.13}$$

in which the coefficients of $\phi_{j-1,s}$ and $\phi_{j+1,s}$ differ. By writing

$$\phi_{j,s} = \gamma^{j} \Psi_{j,s} \tag{3.14}$$

where γ is given by

$$\gamma = \sqrt{\left[\frac{1-2bh}{1+2bh}\right]}, \tag{3.15}$$

equation (3.13) may be written in the symmetric form

$$\alpha \Psi_{j-1,s} - 4\Psi_{j,s} + \alpha\Psi_{j+1,s} + \Psi_{j,s-1} + \Psi_{j,s+1} = h^2 \gamma^{-j} q_{j,s} \tag{3.16}$$

where

$$\alpha = \sqrt{(1-2bh)(1+2bh)} . \tag{3.17}$$

The FFT method may thus be used on the systems of equations generated from (3.16) and the solution to the system (3.13) recovered via (3.14). It is clear that we must choose

$$h < \frac{1}{2 |b|} . \tag{3.18}$$

The above device was considered by Hildebrand (1968, p227), Le Bail (1972), and has been used, for example, by Pickering and Sozou (1979) in solving an elliptic equation arising in a problem in fluid dynamics.

3.3.2   Elliptic equations in cartesian coordinates

In general, equation (3.10) is purely elliptic if $a(y) > 0$ throughout the region of interest and we concentrate here on elliptic problems which are to be solved over a finite rectangular region $0 \leqslant x \leqslant L_x$, $0 \leqslant y \leqslant L_y$ with, in addition to (3.11), conditions which may be of the form

$$u_1(x)\, \frac{\partial \phi}{\partial x} + u_2(x)\phi = u_3(x) \quad \text{on} \quad y = 0$$

$$v_1(x)\, \frac{\partial \phi}{\partial x} + v_2(x)\phi = v_3(x) \quad \text{on} \quad y = L_y \ , \tag{3.19}$$

or, alternatively, ϕ may be periodic in the y direction. The general forms (3.19) include Neumann and Dirichlet conditions as special cases. We note also that the mesh length h_x must be constant whereas h_y may be variable.

The most straightforward problem is probably that arising from the 5-point (second-order) approximation to Poisson's equation with Dirichlet conditions as described in Chapter 1. We consider two further examples of elliptic problems; the first is essentially a formulation exercise and the second a worked example using two different finite-difference schemes. In both these examples Dirichlet conditions are assumed; an illustration of the use of Neumann conditions is given in a later, three-dimensional, problem.

Example 3.3.1

We consider solving Poisson's equation using the standard 9-point approximation given, for example, by Bickley (1948). On a square grid of side h this approximation may be written as

$$\left[\nabla^2 \phi\right]_{j,s} = \frac{4\textstyle\sum_1 + \textstyle\sum_2 - 20\phi_{j,s}}{h^2} - \frac{1}{12} h^2 \nabla^2 q_{j,s} + 0(h^4) \qquad (3.20)$$

where

and

$$\left.\begin{array}{l} \textstyle\sum_1 = \phi_{j-1,s} + \phi_{j+1,s} + \phi_{j,s-1} + \phi_{j,s+1} \, , \\[2ex] \textstyle\sum_2 = \phi_{j-1,s-1} + \phi_{j+1,s-1} + \phi_{j-1,s+1} + \phi_{j+1,s+1} . \end{array}\right] \qquad (3.21)$$

It is left to the reader to deduce that on a typical row of grid (s = constant) the finite difference equations may be written in the form

$$T_1 \underset{\sim}{\Psi}_{s-1} + T_0 \underset{\sim}{\Psi}_s + T_1 \underset{\sim}{\Psi}_{s+1} = \underset{\sim}{Q}_s \qquad (3.22)$$

where T_0 and T_1 are tridiagonal matrices such that

$$T_0 = 4T_1 + 36I. \qquad (3.23)$$

Clearly T_0 and T_1 commute and it is easy to verify that these matrices satisfy all the requirements for FFT methods to be suitable. Pickering (1977) has reported that, with Dirichlet conditions, the method takes typically 28% more time than the corresponding 5-point method but of course there are considerable gains in accuracy in using the 9-point rather than the 5-point method with the same grid size. Houstis et al., (1977) also formulated a 9-point FFT method for the Helmholtz equation and their detailed results indicate that their method is typically 50 times faster than the best 5-point methods for comparable accuracy to be achieved.

Example 3.3.2

Here we consider a problem in which two different types of
approximation are employed and the results obtained from each are
compared for various grid sizes.

We consider solving an equation of the form

$$\frac{\partial^2 \phi}{\partial x^2} + \frac{\partial^2 \phi}{\partial y^2} + c(y) \frac{\partial \phi}{\partial y} = q(x,y) \qquad (3.24)$$

over the unit square $0 \leqslant x, y \leqslant 1$

with

$$c(y) = -\frac{6}{1.1-y} \qquad (3.25)$$

and

$$q(x,y) = \frac{3\sqrt{xy}}{(3-x)(2-y)} , \qquad (3.26)$$

with boundary conditions given by

$$\phi(0,y) = \phi(1,y) = \phi(x,0) = 0 \quad \text{and} \quad \phi(x,1) = 4x(1-x). \quad (3.27)$$

It is a straightforward matter using standard second-order central
difference approximations, to solve the above problem using FFT
techniques. However, it is worth drawing attention at this stage
to one feature of the above problem, namely that near to the
boundary $y = 1$ relatively large values of $c(y)$ (compared with
unity) are generated. Several authors, notably Allen (1962),
Dennis (1960) and Roscoe (1975), have proposed a form of
approximation which may be applied to (3.24) and involves the
exponential function. Such approximations have been found to give

improved results as compared with those obtained using local

polynomial approximations for a class of elliptic equations which

includes (3.24) as a special case and many authors have used such

approximations to produce diagonally dominant systems of linear

equations which are subsequently solved by iterative methods. Some

further references are given by Pickering (1983). For

completeness, we briefly outline the derivation of the scheme

following Allen's approach. Roscoe's more general analysis

produces the same difference equation as Allen's method for

(3.24).

We write (3.24) as

$$\frac{\partial^2 \phi}{\partial x^2} = \chi(x,y), \qquad (3.28)$$

so that

$$\frac{\partial^2 \phi}{\partial y^2} + c(y) \frac{\partial \phi}{\partial y} = q(x,y) - \chi(x,y), \qquad (3.29)$$

and at a mesh point (x_j, y_s) solve (3.29) as if it were an ordinary

differential equation to obtain the local approximation

$$\phi = C_0 e^{-c_s y} + C_1 + (q_{j,s} - \chi_{j,s})y/c_s \qquad (3.30)$$

where C_0 and C_1 are arbitrary constants. We set $y = y_{s-1}$, y_s and

y_{s+1} in turn in (3.30) thus giving expressions for $\phi_{j,s-1}$ $\phi_{j,s}$ and

$\phi_{j,s+1}$ from which C_0 and C_1 can be eliminated. Using a standard polynomial approximation for $\partial^2\phi/\partial x^2$ we may eliminate $\chi_{j,s}$ to obtain for a square mesh of side h, such that $(R+1)h = 1$

$$\phi_{j-1,s} - (2+U_s+V_s)\phi_{j,s} + \phi_{j+1,s} + U_s\phi_{j,s-1} + V_s\phi_{j,s+1} = h^2 q_{j,s} \, , \qquad (3.31)$$

where

$$\left. \begin{array}{l} U_s = c_s h/(e^{c_s h}-1) \\[12pt] V_s = U_s e^{c_s h} \, . \end{array} \right] \qquad (3.32)$$

and

The truncation error of this type of approximation has a leading term $O(h^2)$. We omit the details of how the boundary conditions are incorporated since the manipulations involved are sufficiently illustrated by the earlier discussion and the example of Chapter 1.

Denoting by $\bar\phi_{r,s}$ $r = 1,2,\ldots,R$ the Fourier harmonics corresponding to a vector ψ_s and by $\bar q_{r,s}$, the harmonics generated from the corresponding right hand side vector q_s, it is easy to deduce that each of the R tridiagonal systems of equations for $\bar\phi_{r,s}$ may be written in the form

$$\lambda_{r,1}\,\bar\phi_{r,1} + V_1\,\bar\phi_{r,2} \qquad\qquad = \bar q_{r,1}$$

$$U_s\,\bar\phi_{r,s-1} + \lambda_{r,s}\,\bar\phi_{r,s} + V_s\,\bar\phi_{r,s+1} = \bar q_{r,s} \qquad s = 2,3,\ldots,R-1$$

$$U_R\,\bar\phi_{r,R-1} + \lambda_{r,R}\bar\phi_{r,R} = \bar q_{r,R} \, , \qquad (3.33)$$

where

$$\lambda_{r,s} = -(U_s + V_s) - 4 \sin^2 \frac{r\pi}{2(R+1)} \quad (r,s = 1,2,\ldots,R).(3.34)$$

Since U_s, $V_s > 0$ it is easily seen that system (3.33) is always diagonally dominant and hence the Thomas (tridiagonal) algorithm of Section 1.5 will be stable for this system. This is not the case for the systems of equations equivalent to (3.33) produced by applying central-difference approximations to (3.24). For these equations the Thomas algorithm is formally stable only for $|c_s| < 2/h$.

Equation (3.24) was solved for five different values of R in the range 7 to 127 using both schemes. As R was increased, it was observed that considerably greater changes took place in the computed solutions obtained using the standard central difference scheme than the corresponding solutions obtained using the scheme based on (3.31). Since our equation does not have a closed form analytical solution, for the purposes of comparison the solution based on (3.31) for R = 127 was thus regarded as "accurate" and the values of RMS error and maximum modulus error given in Table 3.1 were calculated on this basis.

It is clear from this Table that, for all R, the RMS error for the exponential scheme is less than the corresponding standard central difference scheme error and furthermore that, as R decreases, the rate of increase of this error measure is less for the exponential scheme than the standard scheme. The maximum modulus error was found to behave in a generally similar fashion, as indicated in Table 3.1. It is relevant to note here that the major feature of the solution is a relatively large positive peak concentrated in the region $0.85 \leqslant y \leqslant 1$, $0.25 \leqslant x \leqslant 0.75$ with a maximum value of 1 at the point $(0.5,1)$. The maximum modulus error was always found to occur in this area and it thus perhaps unrealistic to expect worthwhile results for $R = 7$. The percentage errors represented by the maximum modulus error for the standard scheme and exponential scheme for $R = 15$ are, respectively, of the order of 30% and 4% and corresponding values for $R = 63$ are approximately 0.5% and 0.1%.

	RMS Error		Max. Mod. Error	
R	Standard	Exponential	Standard	Exponential
7	1.35,-2	9.25,-4	3.24,-1	2.18,-2
15	2.31,-3	3.22,-4	1.77,-1	2.39,-2
31	3.15,-4	6.42,-5	6.52,-2	1.28,-2
63	3.99,-5	7.43,-6	1.72,-2	3.12,-3
127	5.79,-6	–	4.94,-3	–

Table 3.1 Values of RMS error and maximum modulus error for both schemes, derived as explained in the text. (Reproduced, with permission, from J. Comp. Phys, (1983), 51, p361, Academic Press Inc.)

R	Max. Modulus Residual	
	Standard	Exponential
7	5.09,-11	1.82,-12
15	2.00,-11	1.46,-11
31	3.27,-11	3.27,-11
63	9.46,-11	8.00,-11
127	1.46,-10	1.46,-10

Table 3.2 Values of maximum modulus residual for both schemes.

The maximum modulus residual, R_M (over all the grid points),
was also calculated and these results are given in Table 3.2.
These results indicate that for the exponential scheme the values
of R_M always decrease as R decreases, whereas for the standard
scheme this is the case as far as R = 15. For this scheme with
R = 7, the value of R_M is roughly twice as large as its value for
R = 31 and about thirty times the value produced by the exponential
scheme. A probable explanation for this behaviour may be
attributed to round-off error propagation effects in the solution
of the tridiagonal systems. As was noted earlier, the Thomas
algorithm applied to (3.33) is always stable whereas for the
polynomial based scheme we require $|c_s| < 2/h$. This condition is
satisfied for all s for R = 127, 63 and 31 but for R = 7 (h = 1/8)
the maximum value of $|c_s|$ is approximately 27. For R = 15

(h = 1/16) the corresponding value is approximately 37.

We conclude this discussion by noting that by running our programs many times we were able to deduce that typically the method based on the exponential scheme took about 10% more computer time than that using the standard scheme. Some increase in execution time for the exponential scheme is not unexpected but it is clear that only R exponentials need to be evaluated since the coefficient c depends only on y.

3.3.3 Parabolic equations

Although the major impetus in the development of rapid partial differential equation solvers has been directed towards elliptic problems, the methods so far described may also be applied to time-dependent problems governed by the general form (3.10) with $a(y) \equiv 0$. Thus we consider a class of linear parabolic equations represented by

$$\frac{\partial^2 \phi}{\partial x^2} + b \frac{\partial \phi}{\partial x} + c(t) \frac{\partial \phi}{\partial t} + d(t)\phi = q(x,t), \qquad (3.35)$$

where $\phi = \phi(x,t)$, $0 \leqslant x \leqslant L_x$, $0 \leqslant t \leqslant \infty$, and an appropriate initial condition of the form

$$\phi(x,0) = u(x), \qquad (3.36)$$

together with any of the suitable boundary conditions (3.11). An initial condition involving a linear relation between ϕ and $\partial\phi/\partial t$ is also appropriate, but the form (3.36) is that which is most frequently met in practice.

In order to illustrate the details of the method we focus our attention on the standard heat-conduction (or diffusion) equation for which $b = d(t) = q(x,t) \equiv 0$ and $c(t) = -1/K^2$, where K is constant, so that

$$\frac{\partial \phi}{\partial t} = K^2 \frac{\partial^2 \phi}{\partial x^2} . \qquad (3.37)$$

Many suitable finite-difference approximations to (3.37) may be found in the literature and we consider in particular a weighted-average scheme which may be written as

$$\frac{\phi_{j,s+1} - \phi_{j,s}}{h_t} = (1-\gamma)K^2 \left[\frac{\phi_{j+1,s} - 2\phi_{j,s} + \phi_{j-1,s}}{h_x^2} \right]$$

$$+ \gamma K^2 \left[\frac{\phi_{j+1,s+1} - 2\phi_{j,s+1} + \phi_{j-1,s+1}}{h_x^2} \right], \qquad (3.38)$$

where $x_j = jh_x$, $t_s = sh_t$, $(R+1)h_x = L_x$ and γ is a parameter such that $0 \leqslant \gamma \leqslant 1$. For $\gamma = 0$ the above formula is the standard explicit approximation to (3.37) whereas for $\gamma = 1$ the approximation is fully implicit and the Crank-Nicolson scheme is obtained by setting $\gamma = 1/2$. It is clear that, in general, this approximation involves nodal ϕ values at six grid points and is a special case of the general form (3.4) discussed earlier.

Equation (3.38) may be rearranged as

$$-\rho_0\gamma \; \phi_{j-1,s+1} + (1+2\rho_0\gamma) \; \phi_{j,s+1} - \rho_0\gamma \; \phi_{j+1,s+1}$$
$$= \rho_0(1-\gamma) \; \phi_{j-1,s} + \left[1-2\rho_0(1-\gamma) \right]\phi_{j,s} + \rho_0(1-\gamma)\phi_{j+1,s}, \qquad (3.39)$$

where

$$\rho_0 = h_t K^2/h_x^2 . \qquad (3.40)$$

In matrix form, assuming for simplicity Dirichlet conditions on $x = 0$, L_x given by (3.11), relation (3.39) may be written as

$$(I-\rho_0\gamma T) \, \psi_{s+1} = [I+\rho_0(1-\gamma)T]\psi_s + \Gamma_s + \Gamma_{s+1}, \qquad (3.41)$$

where T is the R-th order tridiagonal matrix given by

$$T = \begin{bmatrix} -2 & 1 & & & \\ 1 & -2 & 1 & & \\ & & \cdot \; \cdot \; \cdot & & \\ & & 1 & -2 & 1 \\ & & & 1 & -2 \end{bmatrix} \qquad (3.42)$$

Γ_s and Γ_{s+1} are the R component vectors

$$\left. \begin{array}{l} \Gamma_s = \rho_0(1-\gamma)(f_s,0,0,\ldots,0,F_s)^T \\ \Gamma_{s+1} = \rho_0\gamma(f_{s+1},0,0,\ldots,0,F_{s+1})^T \end{array} \right], \qquad (3.43)$$

and ψ_s denotes a vector whose components are $\phi_{j,s}$, $j = 1,2,\ldots,R$. If we consider solving (3.41) for $s = 0,1,2,\ldots,S$ then a suitable step-by-step procedure may be formulated using FFT techniques.

We expand ψ_s in terms of the eigenvectors x_r of T so that

$$\psi_s = \sum_{r=1}^{R} \bar{\phi}_{r,s} x_r , \qquad (3.44)$$

and similarly $\Gamma_s + \Gamma_{s+1}$. Thus we obtain from (3.41)

$$(1-\rho_0\gamma\lambda_r)\bar{\phi}_{r,s+1} = [1+\rho_0(1-\gamma)\lambda_r]\bar{\phi}_{r,s} + \bar{d}_{r,s} \qquad (3.45)$$

where $\bar{d}_{r,s}$ denotes the harmonics associated with $\Gamma_s + \Gamma_{s+1}$

and λ_r is the eigenvalue of T corresponding to $\underset{\sim}{x}_r$, given by

$$\lambda_r = - 4\sin^2 \frac{r\pi}{2(R+1)} . \qquad (3.46)$$

Since $\bar{d}_{r,s}$ is known and $\underset{\sim}{\psi}_0$ is given via (3.36) so that $\bar{\phi}_{r,0}$ is known for $r = 1,2,\ldots,R$, equation (3.45) is an _explicit_ formula which may be used to calculate the required harmonics $\bar{\phi}_{r,s}$, $s = 1,2,\ldots,S$; $r = 1,2,\ldots,R$. Hence, in the usual way, the final solution may be synthesised using (3.44).

It is well-known that the approximation (3.38) is stable for $\rho_0 > 0$ and $1/2 \leqslant \gamma \leqslant 1$, whereas for $0 \leqslant \gamma < 1/2$ we require $\rho_0 \leqslant 1/[2(1-2\gamma)]$. It is left as an exercise for the reader to verify that the above results follow from (3.45) by requiring that

$$\left| 1-\rho_0\gamma\lambda_r \right| \geqslant \left| 1+\rho_0(1-\gamma)\lambda_r \right|. \qquad (3.47)$$

The truncation error, $E(x)$, associated with (3.39) may be written as

$$E(x) = \rho_0 h_x^4 [(\tfrac{1}{2}\gamma)\rho_0 - \tfrac{1}{12}]\phi_{xxxx} + O(h_x^6), \qquad (3.48)$$

where it is implied that ρ_0 is held fixed in a sequence of mesh refinements as h_x is decreased. For $\gamma \neq 1/2$ it follows from (3.48) that the truncation error associated with (3.38) is $O(h_t) + O(h_x^2)$ whereas for the case $\gamma = 1/2$ (the Crank-Nicolson scheme) the corresponding truncation error is $O(h_t^2) + O(h_x^2)$. It is also well-known that this scheme is convergent for $\rho_0 > 0$. A detailed

discussion of the behaviour of the truncation error of the weighted
average approximation may be found, for example, in Hildebrand
(1968, p209) under various assumptions regarding the behaviour of
h_t and h_x as the mesh is refined.

The foregoing illustrative example demonstrates how the FFT
method shows to advantage with implicit formulations to parabolic
equations since the Fourier harmonics may be calculated from an
explicit formula which is unconditionally stable. The methodology
for a more general equation of the form (3.35) and other forms of
approximation follow along similar lines.

3.3.4 Hyperbolic equations

Hyperbolic problems which involve no discontinuities in
initial and/or boundary data may be readily solved by
finite-difference methods. Equation (3.10) is hyperbolic for
$a(y) < 0$ and usually the variable y represents time t. Thus with
$\phi = \phi(x,t)$ suitable initial conditions for such an equation may be
written as

$$\phi(x,0) = u(x) \qquad\qquad (3.49)$$
$$(0 \leqslant x \leqslant L_x)$$
$$\text{and} \qquad \frac{\partial \phi}{\partial t}(x,0) = v(x), \qquad\qquad (3.50)$$

where u(x) and v(x) are prescribed functions and any of the
appropriate boundary conditions (3.11) may be employed. In
contrast with the parabolic case the conditions (3.49) and (3.50)
allow the determination of $\phi(x,t)$ for $t > 0$ and/or $t < 0$.

In view of the discussion of the previous Section for the
parabolic case, we would expect essentially the same advantages for
the FFT method for implicit approximations to hyperbolic equations
of the general form (3.10). Le Bail (1972) considers an explicit
approximation to the wave-equation

$$\frac{\partial^2 \phi}{\partial x^2} = \frac{1}{c^2} \frac{\partial^2 \phi}{\partial t^2} , \quad (t \geqslant 0) \quad (3.51)$$

which is stable and convergent for $\rho_0 \leqslant 1$, where $\rho_0 = ch_t/h_x$, and
formulates the computational procedure for this case. As an
example of an implicit formulation to (3.51) we use the
approximation

$$\frac{\partial^2 \phi}{\partial x^2}j,s \approx \frac{1}{2}[\frac{\partial^2 \phi}{\partial x^2}j,s+1 + \frac{\partial^2 \phi}{\partial x^2}j,s-1], \quad (3.52)$$

so that at a mesh point (x_j, t_s), (3.51) may be approximated by

$$- \rho_0^2 \phi_{j-1,s+1} + 2(1+\rho_0^2) \phi_{j,s+1} - \rho_0^2 \phi_{j+1,s+1} \quad (3.53)$$
$$= 4\phi_{j,s} + \rho_0^2 \phi_{j+1,s-1} - 2(1+\rho_0^2) \phi_{j,s-1} + \rho_0^2 \phi_{j-1,s-1}$$

Hence in matrix form, assuming Dirichlet boundary conditions,
relation (3.53) may be written as

$$(2I - \rho_0^2 T) \, \underset{\sim}{\phi}_{s+1} = 4I \, \underset{\sim}{\phi}_s - (2I - \rho_0^2 T) \, \underset{\sim}{\phi}_{s-1} + \underset{\sim}{\Gamma}_{s-1} + \underset{\sim}{\Gamma}_{s+1}, \quad (3.54)$$

where T is the Rth order matrix given by (3.42) and $\underset{\sim}{\Gamma}_{s-1}$ and $\underset{\sim}{\Gamma}_{s+1}$
are R-component vectors given by

$$\underset{\sim}{\Gamma}_{s\pm1} = \rho_0^2(f_{s\pm1}, 0, \ldots, 0, F_{s\pm1})^T. \quad (3.55)$$

Proceeding in the same way as for the parabolic case we may

obtain from (3.54)

$$(2 - \rho_0^2 \lambda_r)\bar{\phi}_{r,s+1} = 4\bar{\phi}_{r,s} - (2 - \rho_0^2 \lambda_r)\bar{\phi}_{r,s-1} + \bar{d}_{r,s} \quad (3.56)$$

where $\bar{d}_{r,s}$ denote the (known) harmonics associated with the vector $\Gamma_{\sim s-1} + \Gamma_{\sim s+1}$ and λ_r is given by (3.46). For this recurrence relation to be stable we require that the eigenvalues of the amplification matrix

$$\begin{bmatrix} \dfrac{4}{2-\rho_0^2\lambda_r} & -1 \\[2ex] 1 & 0 \end{bmatrix}, \tag{3.57}$$

have modulus $\leqslant 1$, and it may be verified that for $\rho_0 > 0$ these eigenvalues are complex with unit modulus.

For $s = 0$ relation (3.56) presents a practical difficulty since information on $s = 0$ and $s = -1$ is required and a number of procedures may be devised to overcome this problem. One method is to approximate (3.50) by a central-difference formula so that

$$\phi_{j,-1} = \phi_{j,1} - 2h_t v_j$$

and use this relation in (3.53) with $s = 0$ to eliminate "fictitious" values on $s = -1$, giving

$$-\rho_0^2\phi_{j-1,1} + 2(1+\rho_0^2)\phi_{j,1} - \rho_0^2\phi_{j+1,1} \tag{3.58}$$
$$= 2u_j - h_t\left[\rho_0^2 v_{j+1} - 2(1+\rho_0^2)v_j + \rho_0^2 v_{j-1}\right].$$

Using (3.58) for $j = 1,2,\ldots,R$ and setting $\phi_{0,1} = f_1$, $\phi_{R+1,1} = F_1$ leads to a tridiagonal system of the form

$$(2I-\rho_0^2 T)\psi_{\sim 1} = \Gamma_{\sim}, \tag{3.59}$$

which can easily be solved for $\psi_{\sim 1}$. Thus relation (3.56) may be used starting with $s = 1$ since the harmonics $\bar{\phi}_{r,1}$ can be calculated from the solution of (3.59) and $\bar{\phi}_{r,0}$ is obtainable from (3.49). An alternative, slightly simpler, approach would be to use the standard explicit scheme to generate $\phi_{j,1}$.

Other implicit approximations to (3.51) could of course be used; for example, the formula

$$\frac{\partial^2}{\partial x^2} \left[\alpha\phi_{j,s+1} + \alpha\phi_{j,s-1} + (1-2\alpha)\phi_{j,s} \right] \approx \frac{\partial^2\phi_{j,s}}{\partial x^2} = \frac{1}{c^2} \frac{\partial^2\phi_{j,s}}{\partial t^2} \qquad (3.60)$$

gives unrestricted stability for $\alpha > 1/4$, and the general approach for linear hyperbolic equations of the form (3.10) is indicated by the above discussion.

3.3.5 Other coordinate systems and grid configurations

An equation such as the two-dimensional Poisson equation in cylindrical polar coordinates (r,θ,z) has the general form of (3.10) with the appropriate interpretation of the independent variables. Thus, for example, if ϕ is independent of θ we have

$$\frac{\partial^2\phi}{\partial z^2} + \frac{\partial^2\phi}{\partial r^2} + \frac{1}{r} \frac{\partial\phi}{\partial r} = q(r,z), \qquad (3.61)$$

whereas if ϕ is independent of z we may write (for $r \neq 0$)

$$\frac{\partial^2\phi}{\partial \theta^2} + r^2 \frac{\partial^2\phi}{\partial r^2} + r \frac{\partial\phi}{\partial r} = r^2 q(r,\theta). \qquad (3.62)$$

For (3.61) or (3.62) the DFT is employed in the z-direction or θ-direction respectively, and appropriate boundary conditions must be associated with these directions. The general procedures for solving such problems then follow along very similar lines to those already described.

Swartztrauber (1974b) has given details of how the discrete
Poisson equation may be solved on the surface of a sphere using FFT
techniques and Sweet (1973) considered the use of staggered grids
which are particularly applicable to certain types of problems in
fluid mechanics. Pickering (1986) solved the discrete Poisson
equation on a regular hexagonal grid using FFT methods and
demonstrated that this 4th order method is typically 30 times
faster than the 5-point (2nd order) method on the basis of speed
for comparable accuracy.

3.4 Higher order equations

In general, if a finite-difference representation of a linear
partial differential and its associated boundary conditions leads
to a suitable block matrix structure of the form indicated earlier,
then the problem may be solved using FFT techniques.

One particular fourth-order example is the biharmonic equation

$$\frac{\partial^4 \phi}{\partial x^4} + 2 \frac{\partial^4 \phi}{\partial x^2 \partial y^2} + \frac{\partial^4 \phi}{\partial y^4} = q(x,y), \qquad (3.63)$$

which may be simply factorised into two Poisson problems. Equation
(3.63) is elliptic in the sense that the boundary is closed and two
conditions need to be specified at all boundary points. If these
conditions represent values of ϕ and $\nabla^2 \phi$ on the boundary, by
setting

$$\frac{\partial^2 \phi}{\partial x^2} + \frac{\partial^2 \phi}{\partial y^2} = \psi , \qquad (3.64)$$

equation (3.63) becomes

$$\frac{\partial^2 \psi}{\partial x^2} + \frac{\partial^2 \psi}{\partial y^2} = q, \qquad (3.65)$$

with ψ given on the boundary. Hence we successively solve the two
Dirichlet problems represented by (3.65) and (3.64) [the calculated
ψ values from (3.65) forming the right hand sides of (3.64)] until
suitable convergence is achieved.

The more common form of boundary conditions for (3.63) are ϕ
and $\partial\phi/\partial n$ given at boundary points, where n denotes the direction
of the outward normal to the boundary. It is still, of course,
possible to factorise (3.63) into (3.64) and (3.65) but the
boundary values of ψ will also vary from the iteration to
iteration. Suitable values may be constructed in a variety of ways
by utilising the given boundary information and equation (3.64) on
the boundary.   Thom & Apelt (1961) discuss this type of approach
as does Smith (1968,1970). Some other fourth order equations may
be factorised in a similar way, as indicated by Le Bail (1972).

3.5   Three dimensional problems

3.5.1 General considerations

The ideas of the previous Sections can be extended in a
straightforward way to problems in three dimensions. The general
form of partial differential equation in three dimensions which may
be solved with appropriate boundary conditions is given by

$$\frac{\partial^2 \phi}{\partial x^2} + a(z) \frac{\partial^2 \phi}{\partial y^2} + b(z) \frac{\partial^2 \phi}{\partial z^2} + c(z) \frac{\partial \phi}{\partial z} + d(z)\phi = q, \qquad (3.66)$$

where z may represent space or time. An example of such a problem may be found in Williams (1969), who considered the three-dimensional Poisson equation in cylindrical polar coordinates. The Fourier analysis and synthesis employed by Williams differs from the usual FFT method described in Chapters 1 and 2 and is based on a technique devised by Danielson & Lanczos (1942).

In general, for the solution of (3.66), two Fourier transforms are performed, first in the x-direction and then in the y-direction. The appropriate harmonics are found either by solving systems of linear equations (for the elliptic case) or by marching procedures (for the parabolic of hyperbolic cases) and the required solution is obtained by two Fourier syntheses. It is perhaps worth noting here that later (Chapter 6), we shall describe a numerical method for solving linear partial differential equations which involve two space variables and time, by the use of a numerical Laplace transform and FFT techniques.

In order to illustrate the general approach for equations of the form (3.66), in the following Section we formulate the method for Poisson's equation in three-dimensions with Neumann boundary conditions.

3.5.2 Poisson's equation in cartesian coordinates

We consider a function $\phi(x,y,z)$ which satisfies

$$\frac{\partial^2 \phi}{\partial x^2} + \frac{\partial^2 \phi}{\partial y^2} + \frac{\partial^2 \phi}{\partial z^2} = q(x,y,z), \qquad (3.67)$$

in a regular cube $0 \leqslant x,y,z \leqslant L$ and the conditions

$$\frac{\partial \phi}{\partial x}(0,y,z) = e(y,z) \left. \rule{0pt}{12pt} \right]$$

$$\frac{\partial \phi}{\partial x}(L,y,z) = E(y,z) \left. \rule{0pt}{12pt} \right] \quad , \tag{3.68}$$

$$\frac{\partial \phi}{\partial y}(x,0,z) = f(x,z) \left. \rule{0pt}{12pt} \right]$$

$$\frac{\partial \phi}{\partial y}(x,L,z) = F(x,z) \left. \rule{0pt}{12pt} \right] \tag{3.69}$$

and

$$\frac{\partial \phi}{\partial z}(x,y,0) = g(x,y) \left. \rule{0pt}{12pt} \right]$$

$$\frac{\partial \phi}{\partial z}(x,y,L) = G(x,y) \left. \rule{0pt}{12pt} \right] \tag{3.70}$$

Since

$$\int_V \nabla^2 \phi \ dV = \int_S \underset{\sim}{\nabla}\phi \cdot d\underset{\sim}{S},$$

equation (3.67) has a solution if and only if

$$\int_S \frac{\partial \phi}{\partial n} \ dS = \int_V q \ dV, \tag{3.71}$$

where S denotes the surface enclosing the volume V of the cube and $\partial \phi/\partial n$ denotes the derivative of ϕ with respect to the (outward) normal to the surface S. Hence the choice of right hand sides in equations (3.68) (3.69) and (3.70) is not arbitrary but must be such that (3.71) is satisfied. Furthermore a function $\phi(x,y,z)$ which satisfies equation (3.67) and the above conditions is only defined to within an additive constant and this arbitrariness is reflected in the fact that the matrix of coefficients derived from (3.67)-(3.70) is singular.

In order to obtain a numerical solution to such a problem ϕ is usually fixed (arbitrarily) at some chosen internal grid point but, for the method of solution based on FFT techniques, rather than fix a value of ϕ it is simpler to arbitrarily choose one of the harmonics, as indicated in the course of the following discussion.

We employ a regular grid of side h, where $(R+1)h = L$ and define the grid-point coordinates by

$$(x_j, y_s, z_\nu) = (jh, sh, \nu h). \quad j,s,\nu = 0,1,\ldots,R+1.$$

Equation (3.67) is approximated by the standard formula

$$\phi_{j-1,s,\nu} -6\phi_{j,s,\nu} +\phi_{j+1,s,\nu} +\phi_{j,s-1,\nu} +\phi_{j,s+1,\nu} +\phi_{j,s,\nu-1} +\phi_{j,s,\nu+1}$$
$$= h^2 q_{j,s,\nu} \qquad (3.72)$$

and we consider first, the boundary conditions on $x = 0,L$ which we represent by

$$\left.\begin{array}{l} \phi_{1,s,\nu} - \phi_{-1,s,\nu} = 2he_{s,\nu} \\[2mm] \phi_{R+2,s,\nu} - \phi_{R,s,\nu} = 2hE_{s,\nu}, \end{array}\right\} \qquad (3.73)$$

and

respectively. Thus by eliminating fictitious values $\phi_{-1,s,\nu}$ and $\phi_{R+2,s,\nu}$ between (3.72) and (3.73) on a grid row parallel to the x-axis (such that $s,\nu \neq 0,R+1$) our finite difference equations may be written as

$$\underset{\sim}{\psi}_{s-1,\nu} +(A-4I)\underset{\sim}{\psi}_{s,\nu} +\underset{\sim}{\psi}_{s+1,\nu} +\underset{\sim}{\psi}_{s,\nu-1} +\underset{\sim}{\psi}_{s,\nu+1} = \underset{\sim}{Q}_{s,\nu}, \qquad (3.74)$$

where

$$\underset{\sim}{\psi}_{s,\nu} = (\phi_{0,s,\nu}, \phi_{1,s,\nu},\ldots, \phi_{R+1,s,\nu})^T, \qquad (3.75)$$

$$\underset{\sim}{Q}_{s,\nu} = (h^2 q_{0,s,\nu} + 2he_{s,\nu}, h^2 q_{1,s,\nu},\ldots,h^2 q_{R,s,\nu}, h^2 q_{R+1,s,\nu} -2hE_{s,\nu})^T$$

and

$$\qquad (3.76)$$

$$A = \begin{bmatrix} -2 & 2 & & & & \\ 1 & -2 & 1 & & & \\ & & \cdot & & & \\ & & & \cdot & & \\ & & 1 & -2 & 1 \\ & & & 2 & -2 \end{bmatrix} \ . \tag{3.77}$$

The conditions (3.69) lead to the results

$$\left. \begin{array}{c} \underset{\sim}{\psi}_{1,\nu} - \underset{\sim}{\psi}_{-1,\nu} = 2h\underset{\sim}{f}_{\nu} \\[2mm] \underset{\sim}{\psi}_{R+2,\nu} - \underset{\sim}{\psi}_{R,\nu} = 2h\underset{\sim}{F}_{\nu} \end{array} \right\} \tag{3.78}$$

and

where $\underset{\sim}{f}_{\nu}$ and $\underset{\sim}{F}_{\nu}$ denote vectors whose components are respectively

values of $f(x_j, z_\nu)$ and $F(x_j, z_\nu)$, $j = 0, 1, \ldots, R+1$.

Hence from (3.74) and (3.78) we find that

$$\begin{bmatrix} \underset{\sim}{\psi}_{0,\nu-1} \\ \underset{\sim}{\psi}_{1,\nu-1} \\ \cdot \\ \cdot \\ \cdot \\ \underset{\sim}{\psi}_{R+1,\nu-1} \end{bmatrix} + \begin{bmatrix} A-4I & 2I & & & \\ I & A-4I & I & & \\ & \cdot & \cdot & & \\ & & \cdot & \cdot & \cdot \\ & I & A-4I & I \\ & & 2I & A-4I \end{bmatrix} \begin{bmatrix} \underset{\sim}{\psi}_{0,\nu} \\ \underset{\sim}{\psi}_{1,\nu} \\ \cdot \\ \cdot \\ \cdot \\ \underset{\sim}{\psi}_{R+1,\nu} \end{bmatrix}$$

$$+ \begin{bmatrix} \underset{\sim}{\psi}_{0,\nu+1} \\ \underset{\sim}{\psi}_{1,\nu+1} \\ \cdot \\ \cdot \\ \cdot \\ \underset{\sim}{\psi}_{R+1,\nu+1} \end{bmatrix} = \begin{bmatrix} \underset{\sim}{\varrho}^{*}_{0,\nu} \\ \underset{\sim}{\varrho}^{*}_{1,\nu} \\ \cdot \\ \cdot \\ \cdot \\ \underset{\sim}{\varrho}^{*}_{R+1,\nu} \end{bmatrix} \ , \tag{3.79}$$

where

$$Q^*_{0,\nu} = Q_{0,\nu} + 2hf_\nu$$

$$Q^*_{s,\nu} = Q_{s,\nu}, \quad s = 1,2,\ldots,R \qquad \nu = 0,1,\ldots,R+1 \quad (3.80)$$

and

$$Q^*_{R+1,\nu} = Q_{R+1,\nu} - 2hF_\nu$$

Boundary conditions (3.70) may be treated similarly and lead to, for $\nu = 0$ and $\nu = R+1$, respectively

$$
\begin{bmatrix}
A-4I & 2I & & & & \\
I & A-4I & I & & & \\
\cdot & \cdot & \cdot & & & \\
& \cdot & \cdot & \cdot & & \\
& & \cdot & \cdot & \cdot & \\
& & I & A-4I & I \\
& & & 2I & A-4I
\end{bmatrix}
\begin{bmatrix}
\psi_{0,0} \\
\psi_{1,0} \\
\cdot \\
\cdot \\
\cdot \\
\cdot \\
\psi_{R+1,0}
\end{bmatrix}
+ 2
\begin{bmatrix}
\psi_{0,1} \\
\psi_{1,1} \\
\cdot \\
\cdot \\
\cdot \\
\cdot \\
\psi_{R+1,1}
\end{bmatrix}
=
\begin{bmatrix}
Q^*_{0,0} + 2hg_0 \\
Q^*_{1,0} + 2hg_1 \\
\cdot \\
\cdot \\
\cdot \\
\cdot \\
Q^*_{R+1,0} + 2hg_{R+1}
\end{bmatrix}
$$

$$(3.81)$$

and

$$
2
\begin{bmatrix}
\psi_{0,R} \\
\psi_{1,R} \\
\cdot \\
\cdot \\
\cdot \\
\psi_{R+1,R}
\end{bmatrix}
+
\begin{bmatrix}
A-4I & 2I & & & & \\
I & A-4I & I & & & \\
\cdot & \cdot & \cdot & & & \\
& \cdot & \cdot & \cdot & & \\
& & \cdot & \cdot & \cdot & \\
& & I & A-4I & I \\
& & & 2I & A-4I
\end{bmatrix}
\begin{bmatrix}
\psi_{0,R+1} \\
\psi_{1,R+1} \\
\cdot \\
\cdot \\
\cdot \\
\cdot \\
\psi_{R+1,R+1}
\end{bmatrix}
=
\begin{bmatrix}
Q^*_{0,R+1} - 2hG_0 \\
Q^*_{1,R+1} - 2hG_1 \\
\cdot \\
\cdot \\
\cdot \\
\cdot \\
Q^*_{R+1,R+1} - 2hG_{R+1}
\end{bmatrix}
$$

$$(3.82)$$

Thus it is convenient to define

$$
\begin{aligned}
\underset{\sim}{Q}^{**}_{s,0} &= \underset{\sim}{Q}^{*}_{s,0} + 2h\underset{\sim}{g}_s \\
\underset{\sim}{Q}^{**}_{s,\nu} &= \underset{\sim}{Q}^{*}_{s,\nu}, \quad \nu = 1,2,\ldots,R
\end{aligned}
\quad \Biggr] \quad s = 0,1,\ldots,R+1 \qquad (3.83)
$$

and $\quad \underset{\sim}{Q}^{**}_{s,R+1} = \underset{\sim}{Q}^{*}_{s,R+1} - 2h\underset{\sim}{G}_s$

We set

$$
\underset{\sim}{\Phi}_{s,\nu} = \sum_{r=0}^{R+1}{}'' \; \bar{\phi}_{r,s,\nu} \; \underset{\sim}{x}_r \qquad (3.84)
$$

and

$$
\underset{\sim}{Q}^{**}_{s,\nu} = \sum_{r=0}^{R+1}{}'' \; \bar{d}_{r,s,\nu} \; \underset{\sim}{x}_r , \qquad (3.85)
$$

where $\sum''$ denotes that the first and last terms in the summation are halved, as indicated in Appendix 1 Section A1.3, and $\underset{\sim}{x}_r$ denotes an eigenvector of A. Substituting (3.84) and (3.85) into (3.79),(3.81) and (3.82), making use of definition (3.83) and pre-multiplying the resulting equation by $\underset{\sim}{x}_r^T D$, where D is a diagonal matrix of order R+2 with unit elements except that $d_{0,0} = d_{R+1,R+1} = 1/2$, we find that

$$
\begin{bmatrix}
B_r & 2I & & & & \\
I & B_r & I & & & \\
 & \cdot & \cdot & \cdot & & \\
 & & \cdot & \cdot & \cdot & \\
 & & & I & B_r & I \\
 & & & & 2I & B_r
\end{bmatrix}
\begin{bmatrix}
\bar{\phi}_{r,0} \\
\bar{\phi}_{r,1} \\
\cdot \\
\cdot \\
\cdot \\
\bar{\phi}_{r,R+1}
\end{bmatrix}
=
\begin{bmatrix}
\bar{d}_{r,0} \\
\bar{d}_{r,1} \\
\cdot \\
\cdot \\
\cdot \\
\bar{d}_{r,R+1}
\end{bmatrix}
, \quad (3.86)
$$

where $\qquad B_r = A + (\lambda_r - 2)I ,$ $\qquad\qquad (3.87)$

λ_r is an eigenvalue of A given by

$$\lambda_r = -4 \sin^2 \frac{\pi r}{2(R+1)}, \tag{3.88}$$

$$\underset{\sim}{\bar{\phi}}_{r,\nu} = (\bar{\phi}_{r,0,\nu}, \bar{\phi}_{r,1,\nu}, \ldots, \bar{\phi}_{r,R+1,\nu})^T \tag{3.89}$$

and

$$\underset{\sim}{\bar{d}}_{r,\nu} = (\bar{d}_{r,0,\nu}, \bar{d}_{r,1,\nu}, \ldots, \bar{d}_{r,R+1,\nu})^T. \tag{3.90}$$

Equations (3.86) may be decoupled by a second Fourier analysis by writing

$$\underset{\sim}{\bar{\phi}}_{r,\nu} = \sum_{p=0}^{R+1}{}'' \bar{\bar{\phi}}_{p,r,\nu} \underset{\sim}{x}_p \tag{3.91}$$

and

$$\underset{\sim}{\bar{d}}_{r,\nu} = \sum_{p=0}^{R+1}{}'' \bar{\bar{d}}_{p,r,\nu} \underset{\sim}{x}_p, \tag{3.92}$$

which leads to tridiagonal systems of the form

$$\mu_{pr} \bar{\bar{\phi}}_{p,r,0} + 2 \bar{\bar{\phi}}_{p,r,1} = \bar{\bar{d}}_{p,r,0}$$

$$\bar{\bar{\phi}}_{p,r,\nu-1} + \mu_{pr} \bar{\bar{\phi}}_{p,r,\nu} + \bar{\bar{\phi}}_{p,r,\nu+1} = \bar{\bar{d}}_{p,r,\nu} \quad (\nu = 1,2,\ldots,R)$$

$$2 \bar{\bar{\phi}}_{p,r,R} + \mu_{pr} \bar{\bar{\phi}}_{p,r,R+1} = \bar{\bar{d}}_{p,r,R+1} \tag{3.93}$$

where

$$\mu_{pr} = \lambda_p + \lambda_r - 2, \tag{3.94}$$

denotes an eigenvalue of B_r.

Thus, in principle, we generate the vectors $\underset{\sim}{Q}^{**}_{s,\nu}$, defined by (3.83), Fourier analyse these to determine $\bar{d}_{r,s,\nu}$ of (3.85) and hence calculate $\bar{\bar{d}}_{p,r,\nu}$ defined by (3.92). For $r = 0$ we solve

(3.93) first for p = 0 then p = 1 and so on to p = R+1 and repeat this process for all values of r. Hence we may construct $\bar{\underset{\sim}{\phi}}_{r,\nu}$ which is defined by (3.91) and then $\underset{\sim}{\psi}_{s,\nu}$, defined by relation (3.84). It is left as an exercise for the reader to show that the matrix of coefficients of (3.93) is singular for the case r = p = 0. Thus, for this case, one could choose, for example, $\bar{\bar{\phi}}_{0,0,r+1} = 0$. An approximate operations count shows that the number of operations for the complete solution of (3.67) is proportional to $N^3 \log_2 N$ where $N = 2^m$ (=R-1). Wilhelmson et al (1977) give further details and consider particularly the use of Fourier analysis and cyclic reduction for solving the 3-dimensional Poisson problem.

3.6 General separable problems

We consider here solving an equation of the form

$$\frac{\partial}{\partial x}\left(a(x)\frac{\partial\phi}{\partial x}\right) + \frac{\partial}{\partial y}\left(b(y)\frac{\partial\phi}{\partial y}\right) + \phi(x,y) = q(x,y) \quad (3.95)$$

with a(x), b(y) > 0 and ϕ specified on a rectangular boundary. Using a square grid of side h and the approximation

$$h^2 \left(\frac{\partial}{\partial x} a(x) \frac{\partial\phi}{\partial x}\right)_{j,s} = \nabla\left(a_{j+\frac{1}{2}}\Delta\phi_{j,s}\right) + O(h^4) \quad (3.96)$$

$$\simeq a_{j+\frac{1}{2}}\phi_{j+1,s} - (a_{j+\frac{1}{2}}+a_{j-\frac{1}{2}})\phi_{j,s} + a_{j-\frac{1}{2}}\phi_{j-1,s},$$

where Δ and ∇ denote respectively the usual forward and backward difference operators, and similarly for the y-derivatives, it is easy to deduce that the full set of finite-difference equations for this problem may be written in the symmetric block-tridiagonal form

$$\begin{bmatrix} B_{11} & B_{12} & & & \\ B_{12} & B_{22} & B_{23} & & \\ & B_{23} & B_{33} & B_{34} & \\ & & \cdot & \cdot & \cdot \\ & & & B_{RR-1} & B_{RR} \end{bmatrix} \begin{bmatrix} \underset{\sim}{\psi}_1 \\ \underset{\sim}{\psi}_2 \\ \cdot \\ \cdot \\ \underset{\sim}{\psi}_R \end{bmatrix} = \begin{bmatrix} \underset{\sim}{g}_1 \\ \underset{\sim}{g}_2 \\ \cdot \\ \cdot \\ \underset{\sim}{g}_R \end{bmatrix} \qquad (3.97)$$

In these equations

$$B_{ss} = \eta_s I + C , \qquad (3.98)$$

where

$$\eta_s = b_{s+1/2} + b_{s-1/2} - h^2 , \qquad (3.99)$$

I is the unit matrix of order R, C is the symmetric tridiagonal matrix

$$C = \begin{bmatrix} -(a_{1/2}+a_{3/2}) & a_{3/2} & & \\ a_{3/2} & -(a_{3/2}+a_{5/2}) & a_{5/2} & \\ & \cdot & \cdot & \cdot \\ & & a_{R-1/2} & -(a_{R-1/2}+a_{R+1/2}) \end{bmatrix} ,$$

$$(3.100)$$

and

$$B_{ss+1} = b_{s+1/2} I . \qquad (3.101)$$

Since the matrix C is symmetric it possesses a complete set of mutually orthogonal eigenvectors and from (3.98) it is clear that the eigensolutions of <u>all</u> the blocks B_{ss} are easily found once those of C have been determined. These eigensolutions will not, in general, be expressible in terms of suitable trigonometric functions since the elements of C depend on j and thus the FFT methods described in the previous Sections cannot be used at the appropriate stages of matrix decomposition. Otherwise, the solution of (3.97) may be obtained in much the same way as in earlier examples. We note here, however, that fast solvers based on cyclic reduction [Swartztrauber (1974a)] or marching [Bank (1977)] are available for equations of the form (3.95), although the details are complicated.

CHAPTER 4
Cyclic Reduction

4.1 Introduction

In the previous Chapter we considered the use of FFT methods
to solve partial differential equations which, when approximated by
suitable finite difference schemes, lead to special systems of
algebraic equations. In particular, for elliptic problems, the
coefficient matrix usually has a block-tridiagonal structure and a
simple illustration is provided by the symmetric block-tridiagonal
matrix of equations (1.7). In the method of cyclic reduction this
structure is utilised by systematically eliminating sets of
unknowns using successive block eliminations and, for suitable
problems, the sequence of systems of equations thus produced may be
recursively generated. Efficient techniques for solving these
equations and the eliminated equations are presented in subsequent
Sections together with a discussion of the stability of the method.

In order to develop the most straightforward method of cyclic
reduction we consider the system of equations

$$B\underset{\sim}{\psi} = \underset{\sim}{g}, \qquad (4.1)$$

for which B is a real symmetric block-tridiagonal matrix with block-order σ given by

$$
B = \begin{bmatrix}
A & V & & & & \\
V & A & V & & & \\
& \cdot & \cdot & \cdot & & \\
& & \cdot & \cdot & \cdot & \\
& & & V & A & V \\
& & & & V & A
\end{bmatrix} , \tag{4.2}
$$

where the matrices A and V are assumed to commute and are both square matrices of order ρ. The vectors $\underset{\sim}{\psi}$ and $\underset{\sim}{q}$ are assumed to be partitioned as in equations (3.3). For Poisson's equation with Dirichlet conditions the matrix A is given by (1.8) and V is essentially the unit matrix. Here, for notational simplicity, we assume that terms involving known boundary values ($\underset{\sim}{\psi}_0$ and $\underset{\sim}{\psi}_{\sigma+1}$) have been taken to the right-hand side in (4.1) and we concentrate on a matrix B which has the structure (4.2) with $\sigma = n - 1$, where $n = 2^{k+1}$. Problems with Neumann or periodic boundary conditions give rise to relatively minor variations in the form of B and details are given in Buzbee et al (1970).

4.2 The reduction procedure

We begin considering a subset of equations (4.1) which we write as

$$
\begin{aligned}
V\underset{\sim}{\psi}_{s-2} + A\underset{\sim}{\psi}_{s-1} + V\underset{\sim}{\psi}_{s} &= \underset{\sim}{q}_{s-1} \\
V\underset{\sim}{\psi}_{s-1} + A\underset{\sim}{\psi}_{s} + V\underset{\sim}{\psi}_{s+1} &= \underset{\sim}{q}_{s} \\
V\underset{\sim}{\psi}_{s} + A\underset{\sim}{\psi}_{s+1} + V\underset{\sim}{\psi}_{s+2} &= \underset{\sim}{q}_{s+1} .
\end{aligned} \tag{4.3}
$$

By multiplying the sum of the first and last of these equations by V and subtracting A times the second equation we may eliminate $\psi_{s\pm1}$ to obtain

$$V^2\psi_{s-2} + (2V^2 - A^2)\psi_s + V^2\psi_{s+2} = Vg_{s-1} - Ag_s + Vg_{s+1}. \qquad (4.4)$$

Similar block eliminations may be carried out using the first three equations of (4.1) and also the last three equations, hence leading to the reduced system, for s even,

$$
\begin{bmatrix}
2V^2 - A^2 & V^2 & & & \\
V^2 & 2V^2 - A^2 & V^2 & & \\
& \cdot & \cdot & \cdot & \\
& & \cdot & \cdot & \cdot \\
& & V^2 & 2V^2 - A^2 & V^2 \\
& & & V^2 & 2V^2 - A^2
\end{bmatrix}
\begin{bmatrix}
\psi_2 \\
\psi_4 \\
\cdot \\
\cdot \\
\cdot \\
\psi_{n-2}
\end{bmatrix}
=
\begin{bmatrix}
Vg_1 - Ag_2 + Vg_3 \\
Vg_3 - Ag_4 + Vg_5 \\
\cdot \\
\cdot \\
\cdot \\
Vg_{n-3} - Ag_{n-2} + Vg_{n-1}
\end{bmatrix}
$$

$$(4.5)$$

for the $2^k - 1$ vectors ψ_2, ψ_4, ..., ψ_{n-2}.

The equations which have been eliminated may be written as

$$
\begin{bmatrix}
A & 0 & . & . & . & . & 0 \\
0 & A & 0 & . & . & . & 0 \\
. & & & & . & & \\
. & & & & . & & \\
0 & . & . & 0 & A & 0 & \\
0 & & & & 0 & A &
\end{bmatrix}
\begin{bmatrix}
\psi_1 \\
\psi_3 \\
. \\
. \\
. \\
\psi_{n-1}
\end{bmatrix}
=
\begin{bmatrix}
g_1 - V\psi_2 \\
g_3 - V(\psi_2 + \psi_4) \\
. \\
. \\
. \\
g_{n-1} - V\psi_{n-2}
\end{bmatrix} ,
\tag{4.6}
$$

which, in principle, may be solved for the 2^k odd-subscripted ψ_s once the even-subscripted ψ_s have been obtained from (4.5).

For suitable matrices A and V the methods of Chapter 3 may be used to solve (4.5) and the block-diagonal structure of equations (4.6) allows the determination of ψ_s for s odd by solving 2^k independent systems of equations. Before considering this point in more detail, however, we note that since the coefficient matrix in (4.5) has a symmetric block-tridiagonal structure, further odd/even block-reductions of the type just outlined are possible and this approach leads to a recursive procedure known as block cyclic reduction. At any stage the reduction process may be stopped and the reduced set of equations solved by a suitable method such as that described in the previous Chapter. Alternatively, the reduction procedure can be carried out as far as possible (k reductions) until the system corresponding to

(4.5) has only one block. Unfortunately, the reduction procedure based on the straightforward approach described above turns out to be unstable, essentially through an increasing loss of significance in the determination of the right-hand side vectors as the number of reductions increases. This is discussed in more detail in Section 4.4 and an alternative procedure which stabilises the calculation is described in Section 4.5. The eliminated equations at any stage have the block diagonal structure indicated by (4.6), but the blocks rapidly fill as the number of reductions increases and an efficient matrix factorisation technique for solving these equations and also the reduced equations [Bunemann (1969)] is given in Section 4.3.

In order to formulate the general reduction procedure we define

$$A^{(o)} = A, \quad V^{(o)} = V, \quad g_s^{(o)} = g_s, \quad (s = 1, 2, \ldots, n-1) \quad (4.7)$$

where, as before, g_1 and g_{n-1} are assumed to include terms which take account of the known boundary values, and for $\ell = 1, 2, \ldots, k$ we set

$$\left. \begin{aligned} A^{(\ell)} &= 2[V^{(\ell-1)}]^2 - [A^{(\ell-1)}]^2 \\ V^{(\ell)} &= [V^{(\ell-1)}]^2 \\ \text{and} \quad g_s^{(\ell)} &= V^{(\ell-1)}[g_{s-2^{\ell-1}}^{(\ell-1)} + g_{s+2^{\ell-1}}^{(\ell-1)}] - A^{(\ell-1)} g_s^{(\ell-1)} \end{aligned} \right\} \quad (4.8)$$

Thus, at each stage our reduced system of equations is of the form

$$T^{(\ell)}\underset{\sim}{\psi}^{(\ell)} = \underset{\sim}{f}^{(\ell)}, \tag{4.9}$$

where

$$T^{(\ell)} = \begin{bmatrix} A^{(\ell)} & V^{(\ell)} & & & \\ V^{(\ell)} & A^{(\ell)} & V^{(\ell)} & & \\ & \cdot & \cdot & \cdot & \\ & & V^{(\ell)} & A^{(\ell)} & V^{(\ell)} \\ & & & V^{(\ell)} & A^{(\ell)} \end{bmatrix} \tag{4.10}$$

is a block-tridiagonal matrix consisting of $(2^{-\ell}n-1)^2$ blocks,

$$\underset{\sim}{\psi}^{(\ell)} = \begin{bmatrix} \underset{\sim}{\psi}_{1\times 2^{\ell}} \\ \underset{\sim}{\psi}_{2\times 2^{\ell}} \\ \underset{\sim}{\psi}_{3\times 2^{\ell}} \\ \cdot \\ \cdot \\ \cdot \\ \underset{\sim}{\psi}_{n-2^{\ell}} \end{bmatrix} \quad \text{and} \quad \underset{\sim}{f}^{(\ell)} = \begin{bmatrix} \underset{\sim}{g}^{(\ell)}_{1\times 2^{\ell}} \\ \underset{\sim}{g}^{(\ell)}_{2\times 2^{\ell}} \\ \underset{\sim}{g}^{(\ell)}_{3\times 2^{\ell}} \\ \cdot \\ \cdot \\ \cdot \\ \underset{\sim}{g}^{(\ell)}_{n-2^{\ell}} \end{bmatrix} . \tag{4.11}$$

For $\ell = 0$ equations (4.9)-(4.11) are simply the original equations and if the reductions are carried out to level k, equations (4.9) take the form

$$A^{(k)}\underset{\sim}{\psi}_{2^k} = \underset{\sim}{g}_2^{(k)}k. \tag{4.12}$$

The eliminated equations, for $\ell = 1,2,\ldots,k$, may be written as

$$D^{(\ell)}\underset{\sim}{\phi}^{(\ell)} = \underset{\sim}{e}^{(\ell)}, \tag{4.13}$$

where

$$D^{(\ell)} = \begin{bmatrix} A^{(\ell-1)} & & & & \\ & A^{(\ell-1)} & & & \\ & & \cdot & & \\ & & & \cdot & \\ & & & & A^{(\ell-1)} \end{bmatrix} \qquad (4.14)$$

is a block-diagonal matrix which has $(2^{-\ell}n)^2$ blocks,

$$\underset{\sim}{\Phi}^{(\ell)} = \begin{bmatrix} \underset{\sim}{\psi}_{1\times 2^{\ell}-2^{\ell-1}} \\ \underset{\sim}{\psi}_{2\times 2^{\ell}-2^{\ell-1}} \\ \underset{\sim}{\psi}_{3\times 2^{\ell}-2^{\ell-1}} \\ \cdot \\ \cdot \\ \cdot \\ \underset{\sim}{\psi}_{n-2^{\ell-1}} \end{bmatrix} \qquad (4.15)$$

and

$$\underset{\sim}{e}^{(\ell)} = \begin{bmatrix} g^{(\ell-1)}_{1\times 2^{\ell}-2^{\ell-1}} - V^{(\ell-1)}\underset{\sim}{\psi}_{1\times 2^{\ell}} \\ g^{(\ell-1)}_{2\times 2^{\ell}-2^{\ell-1}} - V^{(\ell-1)}(\underset{\sim}{\psi}_{1\times 2^{\ell}} + \underset{\sim}{\psi}_{2\times 2^{\ell}}) \\ g^{(\ell-1)}_{3\times 2^{\ell}-2^{\ell-1}} - V^{(\ell-1)}(\underset{\sim}{\psi}_{2\times 2^{\ell}} + \underset{\sim}{\psi}_{3\times 2^{\ell}}) \\ \cdot \\ \cdot \\ \cdot \\ g^{(\ell-1)}_{n-2^{\ell-1}} - V^{(\ell-1)}\underset{\sim}{\psi}_{n-2^{\ell}} \end{bmatrix} \cdot \qquad (4.16)$$

As indicated earlier, the reduction procedure may be stopped at any stage $0 < \ell < k$ and the resulting equations solved by any suitable method. For small values of ℓ direct solution by say an elimination procedure is practicable but if the matrices A and V are such that equations (4.1) may be solved by the FFT method then equations (4.10) and (4.14) also may be solved by this method. Denoting by $\lambda_i^{(o)}$ an eigenvalue of $A^{(o)}$ and by $\mu_i^{(o)}$ an eigenvalue of $V^{(o)}$, since $A^{(o)}$ and $V^{(o)}$ are assumed to commute it is easy to show that the eigenvalues of $A^{(\ell)}$ and $V^{(\ell)}$, denoted, respectively, by $\lambda_i^{(\ell)}$ and $\mu_i^{(\ell)}$, may be calculated from the relations

$$\left.\begin{aligned} \lambda_i^{(\ell)} &= 2\left(\mu_i^{(\ell-1)}\right)^2 - \left(\lambda_i^{(\ell-1)}\right)^2, \\ \mu_i^{(\ell)} &= \left(\mu_i^{(\ell-1)}\right)^2 \; . \end{aligned}\right\} \tag{4.17}$$

An alternative method, based on matrix factorisation, may be used to solve both the reduced and eliminated equations and this is explained in the following Section.

4.3 Matrix factorisation

We begin by considering the system of equations (4.12), which are obtained after k levels of reduction and, in particular, we examine the structure of the coefficient matrix $A^{(k)}$. From equations (4.8) we note that

$$A^{(1)} = 2\left(V^{(o)}\right)^2 - \left(A^{(o)}\right)^2 = 2V^2 - A^2, \qquad (4.18)$$

so that $A^{(1)}$ is a polynomial of degree 2 in both A and V.

Similarly

$$\left.\begin{aligned} A^{(2)} &= 2\left(V^{(1)}\right)^2 - \left(A^{(1)}\right)^2 \\ &= -2V^4 + 4V^2A^2 - A^4, \\ A^{(3)} &= -2V^8 + 16V^6A^2 - 20V^4A^4 + 8V^2A^6 - A^8 \end{aligned}\right\} \quad (4.19)$$

and so on, from which it follows that, in general, $A^{(\ell)}$ is a matrix polynomial of degree 2^ℓ in both A and V which we may write as

$$A^{(\ell)} = P_{2^\ell}(A,V) = \sum_{\nu=0}^{2^{\ell-1}} c_{2\nu}{}^{(\ell)} A^{2\nu} V^{2^\ell - 2\nu} \qquad (4.20)$$

whose coefficient of A^{2^ℓ} is given by

$$c_{2^\ell}^{(\ell)} = -1 . \qquad (4.21)$$

It is possible to determine the linear factors of (4.20) by the following analysis. We consider the scalar polynomial

$$P_{2^\ell}(a,v) = \sum_{\nu=0}^{2^{\ell-1}} c_{2\nu}^{(\ell)} a^{2\nu} v^{2^\ell - 2\nu}, \qquad (4.22)$$

which satisfies the relation [see equations (4.8)]

$$P_{2^{\ell+1}}(a,v) = 2v^{2^{\ell+1}} - \left[P_{2^\ell}(a,v)\right]^2, \qquad (4.23)$$

and set

$$a/2v = -\cos\theta, \qquad (4.24)$$

so that

$$p_1 = 2v^2 - a^2 = -2v^2 \cos 2\theta,$$

$$p_4 = 2v^4 - p_1^2 = -2v^4 \cos 4\theta,$$

$$p_8 = -2v^8 \cos 8\theta$$

and, in general,

$$p_{2^\ell}(a,v) = -2v^{2^\ell} \cos 2^\ell \theta. \qquad (4.25)$$

Thus, $p_{2^\ell}(a,v) = 0$ for

$$\theta = \theta_\mu^{(\ell)} = \left[\frac{2\mu-1}{2^{\ell+1}}\right] \pi, \ \mu = 1,2,\ldots,2^\ell. \qquad (4.26)$$

Hence, using (4.21), we may express $p_{2^\ell}(a,v)$ as a product of linear

factors in the form

$$p_{2^\ell}(a,v) = -\prod_{\mu=1}^{2^\ell} (a+2v \cos \theta_\mu^{(\ell)}), \qquad (4.27)$$

so that

$$A^{(\ell)} = -\prod_{\mu=1}^{2^\ell} G_\mu^{(\ell)}, \qquad (4.28)$$

where

$$G_\mu^{(\ell)} = A + 2 \cos \theta_\mu^{(\ell)} V. \qquad (4.29)$$

This factorisation enables the solution of equations (4.12) to

be found by solving a sequence of tridiagonal systems, since, from

(4.12) and (4.18), we have

$$\left[\prod_{\mu=1}^{2^k} G_\mu^{(k)}\right] \underset{\sim}{\Psi}_{2^k} = -\underset{\sim}{g}_{2^k}^{(k)}, \qquad (4.30)$$

and setting $\underset{\sim}{z}_1 = -\underset{\sim}{g}_{2^k}^{(k)}$, we may repeatedly solve

$$G_\mu^{(k)} \underset{\sim}{z}_{\mu+1} = \underset{\sim}{z}_\mu, \quad \mu = 1,2,\ldots,2^k \qquad (4.31)$$

thus giving finally

$$\psi_2 k = z_2 k_{+1}. \qquad (4.32)$$

A similar technique may be used to solve the eliminated equations (4.14), with the factorisation given by (4.28), and to compute $g_s^{(\ell)}$ in equations (4.8). An alternative way of computing $A^{(\ell-1)} g_s^{(\ell-1)}$ which appears in (4.8) is given by Buzbee et al (1970) which utilises the recursive nature of the polynomials (4.25). With $p_\mu(a,v)$ defined by

$$p_\mu(a,v) = - 2v^\mu \cos \mu\theta, \qquad (4.33)$$

it is easy to show that

$$\left. \begin{array}{l} p_\mu(a,v) = - ap_{\mu-1}(a,v) - v^2 p_{\mu-2}(a,v), \quad \mu \geqslant 2 \\ p_o(a,v) = - 2, \quad p_1(a,v) = a, \end{array} \right\} \quad (4.34)$$

so that by computing

$$\left. \begin{array}{l} \eta_o = -2g_s^{(\ell-1)}, \quad \eta_1 = Ag_s^{(\ell-1)}, \\ \eta_\mu = - A\eta_{\mu-1} - V^2\eta_{\mu-2}, \quad \mu = 2,3,\ldots,2^{\ell-1} \end{array} \right\} \quad (4.35)$$

we obtain

$$\eta_2\ell_{-1} = A^{(\ell-1)} g_s^{(\ell-1)}. \qquad (4.36)$$

The overall method of solution of equations (4.1) as described above with k levels of reduction is known as the cyclic odd-even reduction and factorisation (CORF) algorithm. Summarising the procedure we see that k reductions are performed to obtain equations (4.12) which are solved by matrix factorisation and this is followed by a sequence of solutions of the eliminated equations (4.13) for ℓ = k,k-1,...,1. Details of the matrix structure and factorisation procedure for Poisson's equation with Neumann or

periodic conditions and several other applications of the technique
may be found in Buzbee et al (1970).

4.4 Stability

Early attempts to use complete reduction based on the
relations derived in the previous Section were failures. The
essential reason was found to be that the procedure for calculating
the right-hand sides of the reduced equations is unstable and we
illustrate this point by considering Poisson's equation on a
regular square grid with Dirichlet boundary conditions, for which
the matrix V is the unit matrix and the matrix A is given by

$$
A = \begin{bmatrix}
-4 & 1 & & & & \\
1 & -4 & 1 & & & \\
& & \cdot & \cdot & \cdot & \\
& & & \cdot & \cdot & \cdot \\
& & & 1 & -4 & 1 \\
& & & & 1 & -4
\end{bmatrix} . \tag{4.37}
$$

We note from equations (4.8) that part of the right-hand side
of the reduced equations involves the calculation of $A^{(\ell-1)} g_s^{(\ell-1)}$
and to investigate the stability of such a computation we consider
ρ-component vectors $\underset{\sim}{v}$ and $\underset{\sim}{y}$ which satisfy

$$
\underset{\sim}{v} = M\underset{\sim}{y}, \tag{4.38}
$$

where M is a given, non-singular, symmetric matrix of order ρ. We
let $\underset{\sim}{\varepsilon}_v$ and $\underset{\sim}{\varepsilon}_y$ denote the errors in $\underset{\sim}{v}$ and $\underset{\sim}{y}$, respectively, so that

$$
\underset{\sim}{v} + \underset{\sim}{\varepsilon}_v = M(\underset{\sim}{y} + \underset{\sim}{\varepsilon}_y)
$$

and hence

$$
\underset{\sim}{\varepsilon}_v = M \underset{\sim}{\varepsilon}_y . \tag{4.39}
$$

Thus,

$$\| \underset{\sim}{\varepsilon}_v \|_2 \; \leqslant \; \| M \|_2 \| \underset{\sim}{\varepsilon}_y \|_2 \, , \tag{4.40}$$

where $\| \underset{\sim}{\varepsilon}_v \|_2$, $\| \underset{\sim}{\varepsilon}_y \|_2$ denote the usual Euclidean norms of $\underset{\sim}{\varepsilon}_v$ and $\underset{\sim}{\varepsilon}_y$, respectively, and $\| M \|_2$ denotes the spectral norm of M. Furthermore, since $\underset{\sim}{y} = M^{-1} \underset{\sim}{v}$ we can deduce that

$$\| \underset{\sim}{v} \|_2 \; \geqslant \; \| M^{-1} \|_2^{-1} \| \underset{\sim}{y} \|_2 \, , \tag{4.41}$$

so that a measure of the relative error in $\underset{\sim}{v}$ in terms of that $\underset{\sim}{y}$ may be obtained from the relation

$$\frac{\| \underset{\sim}{\varepsilon}_v \|_2}{\| \underset{\sim}{v} \|_2} \; \leqslant \; \| M \|_2 \; \| M^{-1} \|_2 \; \frac{\| \underset{\sim}{\varepsilon}_y \|_2}{\| \underset{\sim}{y} \|_2} \; . \tag{4.42}$$

Since M is symmetric, the amplification factor in relation (4.42) may be written as

$$\| M \|_2 \; \| M^{-1} \|_2 \; = \; \left| \frac{\lambda_\rho}{\lambda_1} \right| \, , \tag{4.43}$$

where λ_ρ and λ_1 denote respectively the spectral radii of M and M^{-1}. The right-hand side of (4.43) is the so-called P-condition number of M.

For the matrix A ($= A^{(o)}$) given by (4.37), all the eigenvalues are in the range $(-6, -2)$ and we are interested in the P-condition number of $A^{(\ell-1)}$. We set

$$\lambda_i^{(o)} = 2 \cosh z_i \, , \tag{4.44}$$

so that using (4.17) (with $\mu_i^{(\ell-1)} = 1$) it is easy to deduce that

$$\lambda_i^{(\ell-1)} = -2 \cosh 2^{\ell-1} z_i \, , \tag{4.45}$$

and hence

$$\left|\frac{\lambda_\rho^{(\ell-1)}}{\lambda_1^{(\ell-1)}}\right| = \frac{\cosh 2^{(\ell-1)}z_\rho}{\cosh 2^{(\ell-1)}z_1} \ , \tag{4.46}$$

where $z_1 = \cosh^{-1}(-\lambda_1^{(o)}/2)$, $z_\rho = \cosh^{-1}(-\lambda_\rho^{(o)}/2)$. A reasonable

approximation to relation (4.46) may be written in the form

$$\left|\frac{\lambda_\rho^{(\ell-1)}}{\lambda_1^{(\ell-1)}}\right| \simeq e^{2^{\ell-1}(z_\rho-z_1)}, \tag{4.47}$$

and thus the number of decimal digits, d, lost in the calculation

of $A^{(\ell-1)}\underset{s}{q}^{(\ell-1)}$ may be estimated from the relation

$$2^{\ell-1}(z_\rho-z_1) = d \ \ell n \ 10. \tag{4.48}$$

Values of d as a function of ℓ derived from (4.48) are given in

Table 4.1.

ℓ	d
1	1
2	1.5
3	3
4	6
5	12
6	24
7	48

Table 4.1 **No. of decimal digits lost as a function of reduction level ℓ.**

This Table demonstrates the rapid way in which significance is

lost in the calculation and indicates that, for practical meshes, reduction to level k will lead to a total loss of significance on most present day computers. The level to which unstabilised reduction may be safely performed is clearly machine dependent and typically, for a machine with 11 decimal digit single precision such as an ICL 1906S, the procedure may be carried out to say ℓ = 3 or 4 without impairing the precision of the results required for most practical applications. For machines such as an IBM 360 or PRIME 750 with six decimal digits (single precision), unstable reduction is patently inadvisable for $\ell > 2$ or 3.

A more complete and rigorous stability analysis is provided by Buzbee et al who derive a relation similar to (4.48) for a more general case.

4.5 Bunemann algorithms

Bunemann's variants of the cyclic reduction algorithm are designed specifically to overcome the difficulties in the calculation of the right-hand sides of equations (4.9). We concentrate here on problems for which the matrix V is the unit matrix, and we begin by deriving what is known as "variant 1" of the cyclic reduction algorithm.

From equation (4.4) we recall that after one level of cyclic reduction we have the relation

$$\psi_{s-2} + (2I-A^2)\psi_s + \psi_{s+2} = g_{s-1} + g_{s+1} - Ag_s \equiv g_s^{(1)}, \quad (4.49)$$

for s = 2,4,...,n-2, where n = 2^{k+1} and $\psi_0 = \psi_n = 0$.

Since, from (4.8), $- A = (A^{(1)}-2I)A^{-1}$, we may write

$$g_s^{(1)} = A^{(1)}A^{-1}g_s + g_{s-1} + g_{s+1} - 2A^{-1}g_s, \qquad (4.50)$$

and hence we may define

$$\underset{\sim}{\alpha}_s^{(1)} = A^{-1}g_s$$

$$\underset{\sim}{\beta}_s^{(1)} = g_{s-1} + g_{s+1} - 2\underset{\sim}{\alpha}_s^{(1)} \Bigg] , \qquad (4.51)$$

and

so that

$$g_s^{(1)} = A^{(1)}\underset{\sim}{\alpha}_s^{(1)} + \underset{\sim}{\beta}_s^{(1)}. \qquad (4.52)$$

After ℓ reductions we have

$$g_s^{(\ell)} = g_{s-2^{\ell-1}}^{(\ell-1)} + g_{s+2^{\ell-1}}^{(\ell-1)} - A^{(\ell-1)}g_s^{(\ell-1)}, \qquad (4.53)$$

which we write as

$$g_s^{(\ell)} = A^{(\ell)}\underset{\sim}{\alpha}_s^{(\ell)} + \underset{\sim}{\beta}_s^{(\ell)} . \qquad (4.54)$$

Substituting this relation in (4.53), and noting that $(A^{(\ell-1)})^2 = 2I - A^{(\ell)}$, we obtain an equation which may be expressed in the form of the two relations

$$\underset{\sim}{\alpha}_s^{(\ell)} = \underset{\sim}{\alpha}_s^{(\ell-1)} - (A^{(\ell-1)})^{-1}[\underset{\sim}{\alpha}_{s-2^{\ell-1}}^{(\ell-1)} + \underset{\sim}{\alpha}_{s+2^{\ell-1}}^{(\ell-1)} - \underset{\sim}{\beta}_s^{(\ell-1)}]$$

and $$\underset{\sim}{\beta}_s^{(\ell)} = - 2\underset{\sim}{\alpha}_s^{(\ell)} + \underset{\sim}{\beta}_{s-2^{\ell-1}}^{(\ell-1)} + \underset{\sim}{\beta}_{s+2^{\ell-1}}^{(\ell-1)} , \qquad (4.55)$$

for $s = i \times 2^\ell$, $i = 1,2,\ldots,2^{k+1-\ell}-1$, with

$$\underset{\sim}{\alpha}_o^{(\ell)} = \underset{\sim}{\alpha}_{2^{k+1}}^{(\ell)} = \underset{\sim}{\beta}_o^{(\ell)} = \underset{\sim}{\beta}_{2^{k+1}}^{(\ell)} = 0. \qquad (4.56)$$

For computational purposes the first of (4.55) is written as

$$A^{(\ell-1)}\left(\underset{\sim}{\alpha}_s^{(\ell-1)} - \underset{\sim}{\alpha}_s^{(\ell)}\right) = \underset{\sim}{\alpha}_{s-2}^{(\ell-1)}{}_{2^{\ell-1}} + \underset{\sim}{\alpha}_{s+2}^{(\ell-1)}{}_{2^{\ell-1}} - \underset{\sim}{\beta}_s^{(\ell-1)}, \quad (4.57)$$

which may be solved using matrix factorisation [equation (4.28)] in
the manner previously described.

In order to backsolve we note that we have the relationship

$$\underset{\sim}{\psi}_{s-2^\ell} + A^{(\ell)}\underset{\sim}{\psi}_s + \underset{\sim}{\psi}_{s+2^\ell} = A^{(\ell)}\underset{\sim}{\alpha}_s^{(\ell)} + \underset{\sim}{\beta}_s^{(\ell)}, \quad (4.58)$$

where

$$\underset{\sim}{\alpha}_s^{(o)} = 0 \quad \text{for} \quad s = 0,1,\ldots,2^{k+1},$$

and

$$\underset{\sim}{\beta}_s^{(o)} = \underset{\sim}{g}_s \quad \text{for} \quad s = 1,2,\ldots,2^{k+1} - 1, \quad (4.59)$$

and we rewrite (4.58) in the form

$$A^{(\ell)}\left(\underset{\sim}{\psi}_s - \underset{\sim}{\alpha}_s^{(\ell)}\right) = \underset{\sim}{\beta}_s^{(\ell)} - \left(\underset{\sim}{\psi}_{s-2^\ell} + \underset{\sim}{\psi}_{s+2^\ell}\right), \quad (4.60)$$

in order to solve using the matrix factorisation procedure. Thus
the Bunemann algorithm, variant 1, may be summarised as follows:

(i) Compute the sequences $\{\underset{\sim}{\alpha}_s^{(\ell)}\}$, $\{\underset{\sim}{\beta}_s^{(\ell)}\}$ using equations (4.55)
for $\ell = 1,2,\ldots,k$ with initial values given by (4.59).

(ii) Use relation (4.60) to backsolve for

$s = 2^\ell$, $3\times 2^\ell$, $5\times 2^\ell$, $\ldots,2^{k+1} - 2^\ell$, noting that $\underset{\sim}{\psi}_o = \underset{\sim}{\psi}_{2^{k+1}} = 0$.

It should be clear to the reader that the solution values
required at any stage on the right-hand side of (4.60) are
naturally available as a result of the calculations of previous
stages (it may be instructive to write out the algorithm for k = 2

or 3). The g_s can be overwritten by the $\beta_s^{(\ell)}$ which in turn can be overwritten by the solution $\underset{\sim}{\psi}_s$, but additional storage is required for the vectors $\underset{\sim}{\alpha}_s^{(\ell)}$. This disadvantage may be overcome, at the cost of some increased computational effort, by eliminating the sequence $\underset{\sim}{\alpha}_s^{(\ell)}$ from equations (4.55) and the resulting procedure is known as Bunemann's "variant 2" of the cyclic reduction algorithm.

To derive this procedure we note from the second of equations (4.55) that

$$\underset{\sim}{\alpha}_s^{(\ell)} = \frac{1}{2} \left(\beta_{s-2h}^{(\ell-1)} + \beta_{s+2h}^{(\ell-1)} - \beta_s^{(\ell)} \right) \tag{4.61}$$

where $h = 2^{\ell-2}$, and substituting this relation in the first of (4.55), after some algebra, we find that

$$\beta_s^{(\ell)} = \beta_{s-2h}^{(\ell-1)} - \beta_{s-h}^{(\ell-2)} + \beta_s^{(\ell-1)} - \beta_{s+h}^{(\ell-2)} + \beta_{s+2h}^{(\ell-1)}$$
$$+ \left(A^{(\ell-1)} \right)^{-1} \left[\beta_{s-3h}^{(\ell-2)} - \beta_{s-2h}^{(\ell-1)} + \beta_{s-h}^{(\ell-2)} - 2\beta_s^{(\ell-1)} + \beta_{s+h}^{(\ell-2)} \right.$$
$$\left. - \beta_{s+2h}^{(\ell-1)} + \beta_{s+3h}^{(\ell-2)} \right] \tag{4.62}$$

for $s = 2^\ell,\ 2 \times 2^\ell, \ldots, 2^{k+1} - 2^\ell$ and $\ell = 2,3,\ldots,k$ with

$$\beta_s^{(o)} = g_s, \text{ for } s = 1,2,\ldots,2^{k+1} - 1, \tag{4.63}$$

and $\beta_s^{(1)} = \beta_{s-1}^{(o)} + \beta_{s+1}^{(o)} - 2A^{-1} \beta_s^{(o)}$ for $s = 2,4,\ldots,2^{k+1} - 2$

The appropriate relation for back substution is obtained by using (4.60) in the form

$$A^{(\ell)} \left[\underset{\sim}{\psi}_s - \frac{1}{2} \left(\beta_{s-2h}^{(\ell-1)} + \beta_{s+2h}^{(\ell-1)} - \beta_s^{(\ell)} \right) \right] = \beta_s^{(\ell)} - \left(\underset{\sim}{\psi}_{s-4h} + \underset{\sim}{\psi}_{s+4h} \right), \tag{4.64}$$

which reduces to the original equations ($\ell = 0$) if we define

$$\beta_{s-1/2}^{(-1)} + \beta_{s+1/2}^{(-1)} - \beta_s^{(o)} = 0.$$

Hence the Bunemann algorithm, variant 2, may be summarised as

(i) Compute the sequence $\beta_s^{(\ell)}$ given by (4.63) and (4.62) for $\ell = 0,1,2,\ldots,k$.

(ii) Backsolve using (4.64) for $\ell = k,k-1,\ldots,0$ and $s = 2^\ell, 3 \times 2^\ell, \ldots, 2^{k+1} - 2^\ell$ with

$$\psi_o = \psi_{2^{k+1}} = \beta_{s-1/2}^{(-1)} + \beta_{s+1/2}^{(-1)} - \beta_s^{(o)} = 0.$$

The extra work required as compared with the first variant amounts to approximately twice as many additions.

A rigorous stability analysis for these algorithms essentially demonstrates that significance is only lost in the calculation of terms which are genuinely small compared with the solution. Details may be found in Buzbee et al (1970), and Swartztrauber (1977) shows that for an N × N grid the asymptotic operation count for the solution of Poisson's equation using variant 2 described above, is $3N^2 \log_2 N$.

4.6 More general cyclic reduction algorithms

Some modifications and extensions to the algorithms just described have been devised. For example, we recall that in the preceding Sections it was necessary for the number of blocks in the direction of reduction to be odd-valued of the form $2^{k+1} - 1$. This restriction has been overcome by Sweet (1977) who described a generalised form of the Buneman algorithms which requires

$O(N^2 \log_2 N)$ operations. Swartztrauber (1974) has successfully generalised the stabilised cyclic reduction procedure so as to be applicable to the general separable equation briefly considered in Chapter 3. The details of both these algorithms are complicated and interested readers are referred to the above articles.

4.7 The FACR(ℓ) method: optimised reduction

As already indicated the cyclic reduction procedure may be stopped at any level $\ell < k$. For suitable problems which are Poisson-like in one variable, the reduced equations may be solved by FFT and the eliminated equations may be solved by the matrix factorisation technique. This combination of Fourier Analysis and ℓ levels of Cyclic Reduction is usually called the FACR(ℓ) algorithm and the question naturally arises,is there an optimum value for the parameter ℓ?

Hockney's algorithm (1965, 1970) halves the number of Fourier transforms required by combining the FFT method with one step of the unstabilised cyclic reduction procedure. Temperton (1979) considers techniques for reducing the length of the Fourier transforms by performing one step of a reduction procedure in the same direction as that in which the Fourier transform is employed and also describes a single-step "diagonal" cyclic reduction procedure which has applications to solving Poisson's equation over an octagonal region [Temperton (1978)]. Very detailed operation counts for these methods and the FACR(ℓ) method (Bunemann variant 1) are given by Temperton (1979, 1980), and Hockney (1970) also gives an operation count for the FACR(ℓ) method. This differs in detail from that given by Temperton, essentially as a result of using different techniques for solving the tridiagonal systems and variations in the implementation of the FFT. Furthermore, an asymptotic operation count for the FACR(ℓ) method using Bunemann's algorithm (variant 2) is given by Swartztrauber (1977) and, for simplicity, we make use of this result here.

For an $N \times N$ grid, defining an operation as a multiplication or division together with an addition or subtraction, Swartztrauber's asymptotic count, $C(\ell)$, for the FACR(ℓ) method may be written as

$$C(\ell) = 3N^2\ell + 2^{1-\ell}N^2\log_2 N. \qquad (4.65)$$

The first term arises from the work required in the reduction and expansion stages (scalar cyclic reduction was used to solve the tridiagonal systems) and the second term arises from the Fourier analysis and synthesis computations. Clearly, as ℓ increases the first term increases and the second decreases, which indicates the existence of a minimum value for C at some value $\ell = \ell*$. Regarding C as a continuous function of ℓ it follows that

$$\ell* = \log_2\log_2 N + \log_2 \left(\frac{2}{3}\ell n2\right) \simeq \log_2\log_2 N - 1 \qquad (4.66)$$

and
$$C(\ell*) \simeq 3N^2\log_2\log_2 N. \qquad (4.67)$$

Values of $\ell*$ calculated from (4.66) are shown in Table 4.2 and values of the asymptotic operation count per grid point, $C(\ell)/N^2$, for $N = 128$ are given in Table 4.3.

$\ell*$	N
1.6	64
1.8	128
2	256
2.2	512
2.3	1024

$C(\ell)/N^2$	ℓ
14	0
10	1
10	2
9	3
13	4

Table 4.2 Values of

$\ell*$ from (4.66).

Table 4.3 Values of

$C(\ell)/N^2$, for N = 128.

Clearly this fairly crude analysis suggests that, for example, for N = 128, $\ell*$ should be chosen as either 2 or 3 and, in practice, Swartztrauber found that $\ell* = 2$ for N = 64 or 128 whereas $\ell* = 3$ for the larger values of N used. Similar theoretical and practical results were obtained by Temperton (1980). Furthermore, precise operation counts indicate that the asymptotic count per grid point is an underestimate by typically a factor of about 2.

It should be fairly apparent to the reader that writing efficient cyclic reduction and FACR(ℓ) ($\ell>0$) programs is a non-trivial undertaking and fortunately a number of (mainly Fortran) programs is available from the above mentioned authors or from program libraries such as the CPC Program Library, Queen's University, Belfast. For example, Christiansen and Hockney (1971) provide a Fortran program for the solution of the two-dimensional Poisson equation in cartesian coordinates and Hughes (1971)

considers the case of cylindrical coordinates. Programming for the
problems and method of Chapter 3 $\left[\text{FACR}(0)\right]$ is, by comparison,
relatively straightforward particularly if as is commonly the case,
a real or complex FFT package is available.

CHAPTER 5
Irregular Regions

5.1. Introduction

In many engineering and scientific calculations the solution
domain does not have a simple geometry and the methods discussed in
the preceding Chapters are inapplicable. In this Chapter therefore
we concentrate on a general method which allows, at some cost in
terms of computational effort, the use of the fast direct methods
of the previous Chapters for solving linear problems with
complicated boundaries. The general approach [see, for example,
Angel (1968)] is to imbed the 'complicated' problem in a simpler,
usually rectangular, domain and by a suitable procedure extract the
solution of the required problem. In this way the equation to be
solved is unchanged and the grid may be chosen as required. We
examine the method from two, slightly different points of view. In
the first we show how a linear combination of suitable solutions
may be determined which satisfies the conditions on the distorted
boundary, assuming that this boundary passes through grid points,
or at least that the boundary is sufficiently well approximated in
this sense. This approach is shown to be equivalent to suitable

111

adjustment of the value of the right-hand side of the equation at appropriate boundary points. The second approach looks at the problem purely from the point of view of linear algebra and establishes an algorithm which formally relates the solutions of the 'easy' and 'difficult' linear equations which respectively approximate the rectangular and imbedded problems. As will be seen, both approaches outlined above necessitate the generation and solution of the so-called 'capacitance-matrix' equations as part of the solution procedure and the following Sections are intended to give the basic details of the method. Some recent developments are outlined in the last Section of this Chapter.

Another approach, for two-dimensional or axisymmetric problems, is to transform a complicated region into, say, a rectangular one using numerical conformal mapping techniques. Many examples of such problems can be found in the literature and Thom and Apelt (1961) and Menikoff and Zemach (1980) provide considerable detail. Challis and Burley (1982) utilise FFT techniques in a method which maps a region which is rectangular, except for one arbitrarily-shaped side, into a rectangle. These authors approximate the distorted side by a finite Fourier sine series and consequently make use of FFT techniques in solving their transformation equations. One difficulty with transformation methods is that resolution can be lost if a region is highly distorted; that is, the choice of a simple, regular, grid in the transformed plane can give rise to a highly non-uniform density of internal grid points in the distorted region [see, for example,

Mobley and Stewart (1980)]. Furthermore, with transformation methods the equation to be solved and the boundary conditions, will, in general be different in the different planes and this can influence the choice of solving technique. These matters are not considered here and interested readers are referred to the literature.

5.2. Linear combinations of solutions: unit source method

We begin by considering a very simple two-dimensional problem in which ϕ satisfies the linear elliptic equation

$$L\phi = \zeta, \qquad (5.1)$$

over the region R with boundary Γ consisting of ABCDEFA as shown in Figure 5.1, and where the region R is imbedded in the rectangular region R' with boundary Γ' consisting of ABD'FA. For illustrative purposes we assume here that ϕ is given on the boundary Γ' and that, for the problem we really wish to solve, ϕ is specified on the boundary Γ. For the grid shown in Figure 5.1, the set of grid points on Γ consists of the grid points on EFABC and the single point D. A more general situation will be considered later.

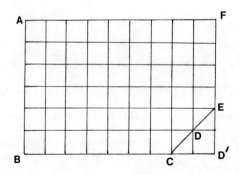

Figure 5.1

To solve this problem we may proceed as follows:

Solve

$$L\phi^{(0)} = \zeta^{(0)}, \tag{5.2}$$

with

$$\phi^{(0)} = g \quad \text{on } \Gamma', \tag{5.3}$$

where $\zeta^{(0)} = \zeta$ except that $\zeta_D^{(0)} = 0$.

Solve

$$L\phi^{(1)} = \zeta^{(1)}, \tag{5.4}$$

with

$$\phi^{(1)} = 0 \quad \text{on } \Gamma', \tag{5.5}$$

where $\zeta^{(1)} = 0$ except that $\zeta_D^{(1)} = 1$. Form the linear combination $\phi = \phi^{(0)} + a\,\phi^{(1)}$, which clearly satisfies $\phi = g$ on Γ' and the equation

$$L\phi = L(\phi^{(0)} + a\,\phi^{(1)}) = \zeta^{(0)} + a\,\zeta^{(1)}, \tag{5.6}$$

whose right hand side is simply ζ at the grid points in R. The constant a is chosen to satisfy the requirement

$$\phi_D = \phi_D^{(0)} + a\,\phi_D^{(1)}. \tag{5.7}$$

We note also that the above analysis implies that if we were to solve (5.1) in R' with the right-hand side modified so that $\zeta_D = a$, our solution would automatically satisfy the requirement at the point D. The following numerical example illustrates the technique for this particular case.

Example 5.2.1

Consider solving Poisson's equation

$$\nabla^2 \phi = 2, \tag{5.8}$$

over the region ABCDEFA shown in Figure 5.2

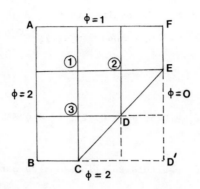

Figure 5.2

which is imbedded in the square ABD'FA on whose boundaries ϕ is given as indicated, and where we require $\phi_D = 1$. Using the usual 5-point discretisation to the Laplacian and assuming, for simplicity, that the grid size, h, is unity the linear equations satisfied by the nodal $\phi^{(0)}$ values are

$$\begin{bmatrix} -4 & 1 & 1 & 0 \\ 1 & -4 & 0 & 1 \\ 1 & 0 & -4 & 1 \\ 0 & 1 & 1 & -4 \end{bmatrix} \begin{bmatrix} \phi_1^{(0)} \\ \phi_2^{(0)} \\ \phi_3^{(0)} \\ \phi_D^{(0)} \end{bmatrix} = \begin{bmatrix} -1 \\ 1 \\ -2 \\ -2 \end{bmatrix}, \tag{5.9}$$

which may be solved to give (correct to 4 decimal places)

$$\phi_1^{(0)} = 0.4583, \quad \phi_2^{(0)} = 0.0417, \quad \phi_3^{(0)} = 0.7917, \quad \phi_D^{(0)} = 0.7083. \tag{5.10}$$

The equations for the nodal $\phi^{(1)}$ values are similar to (5.9) except that the right-hand side vector is $(0, 0, 0, 1)^T$ and thus we obtain

$$\phi_1^{(1)} = -0.0417, \ \phi_2^{(1)} = \phi_3^{(1)} = -0.0833, \ \phi_D^{(1)} = -0.2917. \quad (5.11)$$

Hence the condition $\phi_D = 1$ requires that

$$1 = 0.7083 - 0.2917a,$$

so that $a = -1$ and

$$\phi_1 = 0.5, \ \phi_2 = 0.125, \ \phi_3 = 0.875 \ . \quad (5.12)$$

It may be verified that this solution is also obtained if we solve (5.9) with right-hand side vector $(-1, 1, -2, -3)^T$ which results from setting $\zeta_D = a = -1$, rather than zero.

Although this very simple example clearly could be solved without recourse to such techniques, the above approach (which is essentially a discrete form of Green's function) can be generalised to allow for an arbitrary number of grid points such as the point D. Furthermore, it is clear the boundaries Γ and Γ' need not have any grid points in common.

Thus we consider solving (5.1) in a region R which is imbedded in a rectangular region R' and is such that the boundary Γ of R has p grid points which are not on Γ'. At these points we assume that we require $\phi = \phi_k$, k = 1,2,...,p. For some problems it may be necessary to choose boundary conditions at points on Γ' which are not on Γ and also extend the definition of the right-hand side of (5.1) into the rectangular region. This situation clearly arises if the problem we actually wish to solve is posed in a closed region R which lies wholly within R' so that Γ and Γ' have no common grid points and thus, for example, ϕ may be chosen arbitrarily on Γ' and ζ may be similarly extended. For other problems the region R' and the boundary conditions on Γ' form part of the problem specification.

For simplicity, we assume here that values of ϕ are available on Γ' for which the solution procedure is given below. The extension to other suitable boundary conditions on Γ' may be formulated along similar lines.

Solve

$$L\phi^{(0)} = \zeta^{(0)} \tag{5.13}$$

with
$$\phi^{(0)} = g \quad \text{on } \Gamma', \tag{5.14}$$

where $\zeta^{(0)} = \zeta$ except that $\zeta_k^{(0)} = 0$, k = 1,2,...,p.

Solve, for $k = 1, 2, \ldots, p$,

$$L\phi^{(k)} = \zeta^{(k)} \tag{5.15}$$

with
$$\phi^{(k)} = 0 \quad \text{on} \quad \Gamma', \tag{5.16}$$

where $\zeta^{(k)} = 0$ except that $\zeta_k^{(k)} = 1$. Form the linear combination

$$\phi = \phi^{(0)} + a_1 \phi^{(1)} + a_2 \phi^{(2)} + \ldots + a_p \phi^{(p)}, \tag{5.17}$$

where the constants a_k, $k = 1, 2, \ldots, p$ must satisfy the linear equations

$$
\begin{bmatrix}
\phi_1^{(1)} & \phi_1^{(2)} & \cdots & \phi_1^{(p)} \\
\phi_2^{(1)} & \phi_2^{(2)} & \cdots & \phi_2^{(p)} \\
\vdots & \vdots & & \vdots \\
\phi_p^{(1)} & \phi_p^{(2)} & \cdots & \phi_p^{(p)}
\end{bmatrix}
\begin{bmatrix}
a_1 \\
a_2 \\
\vdots \\
a_p
\end{bmatrix}
=
\begin{bmatrix}
\phi_1 - \phi_1^{(0)} \\
\phi_2 - \phi_2^{(0)} \\
\vdots \\
\phi_p - \phi_p^{(0)}
\end{bmatrix}, \tag{5.18}
$$

in order that $\phi = \phi_k$ at the p relevant grid points on Γ. The notation $\phi_j^{(k)}$ denotes the value of solution $\phi^{(k)}$ at the j-th such point.

The matrix

$$
C = \begin{bmatrix}
\phi_1^{(1)} & \phi_2^{(2)} & \cdots & \phi_1^{(p)} \\
\phi_2^{(1)} & \phi_2^{(2)} & \cdots & \phi_2^{(p)} \\
\cdot & & & \\
\cdot & & & \\
\phi_p^{(1)} & \phi_p^{(2)} & \cdots & \phi_p^{(p)}
\end{bmatrix}^{-1} \qquad (5.19)
$$

is called the 'capacity' (or 'capacitance') matrix since, for
electrostatic problems, ϕ represents the electric potential and the
constants a_k represent the necessary corrections to the charges at
the relevant points in order to obtain the required potentials ϕ_k
at these points. The use of a capacity matrix as a computational
tool was probably first demonstrated by Hockney (1968).

In practice the complete set of nodal $\phi^{(k)}$ values for all grid
points in R' is not stored. Instead, once (5.18) has been solved
for a_k, $k = 1,2,\ldots,p$, the 'easy' problem is solved again in R'
with the correct right-hand sides, that is, we solve

$$L\phi = \overline{\zeta} , \qquad (5.20)$$

with $\qquad \phi = g$ on Γ' , $\qquad (5.21)$

where $\overline{\zeta} = \zeta$, except that $\overline{\zeta}_k = a_k$, $k = 1,2,\ldots,p$. Thus, once the
capacity matrix has been calculated and the corrections determined
we see that the solution for an irregular region can be obtained by
solving two suitably chosen problems in a rectangle, namely (5.13)
and (5.20), and, furthermore, a set of problems (5.1) with different
right-hand sides ζ may be solved using the same capacity matrix.

5.3. Discussion

Hockney (1970, 1978) has used the above approach together with the FACR(1) algorithm to produce a program which is particularly useful in the study of field-effect transistors. The region Γ' is rectangular on whose sides any suitable (separable) boundary conditions for the FACR(1) algorithm may be employed. Extra boundaries (electrodes) may be introduced which may be totally within Γ' or may have some points in common with Γ', including parts of the boundaries x = constant (the Fourier transform is employed, as usual, in the x-direction). Each electrode is represented by a straight line segment and a constant charge is placed on the mesh points nearest to the line so that the potential is correct at such mesh points in an average sense. Hockney's program can thus be utilised for solving the Poisson problem in an arbitrary region R with Dirichlet conditions and essentially gives a first-order method of solution of such problems.

The capacitance matrices generated by the above technique turn out to be symmetric and positive definite and thus advantage can be taken of the Cholesky algorithm for the solution of (5.18). The singular cases of Neumann and periodic conditions on Γ' are also treated by Hockney who also gives detailed operation counts and execution times for illustrative problems. For example, for a test problem involving 100 boundary and 100 internal electrodes for various size grids, Hockney's results show that typically, the computer time required to calculate the capacity matrix is about 50 times the time to solve the problem once the capacity matrix is determined, a factor which is about half the number of internal electrodes. (Boundary electrodes are incorporated in Hockney's algorithm in a more economical way than internal electrodes, loc. cit). Furthermore, the time to form C by numerical inversion was found to be small compared with the time required to generate C^{-1} and, for a 128x128 grid, about 3 decimal digits of precision were lost due to the accumulation of rounding error on an IBM 360/91.

Widlund (1972) has shown how the p solutions of (5.15) can be avoided if the boundary conditions on the rectangle Γ' are doubly-periodic. In this case the relevant matrix is a circulant and a knowledge of one column therefore suffices to determine the entire matrix. Furthermore, for problems with long electrodes, Matte and Lafrance (1977) use a 'clustering' technique so as to reduce the order of the capacitance matrix and, in aerodynamic calculations, Martin (1974) employed a method similar to Hockney's together with an interpolation procedure to allow for the situation where the boundary Γ does not pass through grid points.

In the following Section we re-examine the general imbedding procedure with a view to establishing a precise relation between the discrete solutions of the rectangular and imbedded problems.

5.4. Matrix formulation

Suppose that the full set of n linear equations representing (5.1) in a rectangular region R' with boundary Γ' (Figure 5.3) may be written as

$$B \underset{\sim}{z} = \underset{\sim}{w}, \qquad (5.22)$$

and that this system can easily be solved. The problem we really wish to solve is (5.1) in the region R with boundary Γ with Dirichlet data on Γ and in this region the linear equations representing the differential equation are the same as those in R' except at the special points marked ● in Figure 5.3. Thus, supposing there are p such points, if we extend the definition of this problem into R', treating the points x as regular grid points, we obtain a system of   linear equations

$$M \underset{\sim}{\Phi} = \underset{\sim}{b} , \qquad (5.23)$$

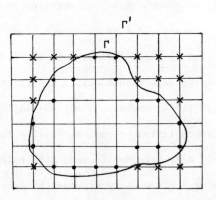

Fig 5.3

in which n − p equations will be the same as n − p equations of system (5.22). We assume that M is non-singular. For simplicity,

but without loss of generality, we shall assume that the first p
rows of B and M differ and, of course, this also applies to the
first p-components of $\underset{\sim}{w}$ and $\underset{\sim}{b}$. Clearly this could be achieved by a
suitable ordering of grid points but in practice the reordering of
rows should be done implicly by indexing since the direct methods
used to solve (5.22) require that B has a certain structure which,
in general, will be different for different orderings. Thus
systems (5.22) and (5.23) may be written in the similarly
partitioned forms

$$B \, \underset{\sim}{z} = \begin{bmatrix} B_1 \\ M_2 \end{bmatrix} \underset{\sim}{z} = \begin{bmatrix} \underset{\sim}{w}_1 \\ \underset{\sim}{b}_2 \end{bmatrix} = \underset{\sim}{w} \tag{5.24}$$

and

$$M \, \underset{\sim}{\Phi} = \begin{bmatrix} M_1 \\ M_2 \end{bmatrix} \underset{\sim}{\Phi} = \begin{bmatrix} \underset{\sim}{b}_1 \\ \underset{\sim}{b}_2 \end{bmatrix} = \underset{\sim}{b} \quad , \tag{5.25}$$

where B_1 and M_1 are p x n matrices, M_2 is an (n-p) x n matrix, $\underset{\sim}{w}_1$
and $\underset{\sim}{b}_1$ are p-component vectors and $\underset{\sim}{b}_2$ has n-p components.

We concentrate here on problems for which the matrix B is
non-singular. An algorithm for the singular case is given in
Buzbee et al (1971), and in both cases the relevant capacitance
matrix is shown to be non-singular. Furthermore, the method of
solution which follows can be shown to be valid whenever the system
(5.23) is consistent.

We begin by noting that we may write

$$M = B + FG \qquad (5.26)$$

where F and G are respectively nxp and pxn matrices given by

$$F = \overline{W} = \begin{bmatrix} W \\ \underset{\sim}{0} \end{bmatrix} \qquad (5.27)$$

and

$$G = W^{-1}(M_1 - B_1) , \qquad (5.28)$$

and W is an arbitrary non-singular pxp matrix, so that W = I is a suitable choice. The Woodbury formula [see, for example, Householder (1964) pp. 123-124] for the inverse of (5.26) may be written as

$$M^{-1} = (B + FG)^{-1} = B^{-1} \left[I - F(I + GB^{-1}F)^{-1}GB^{-1} \right] , \quad (5.29)$$

and if we define the pxp matric $\overline{C}$ as

$$\overline{C} = I + GB^{-1}F , \qquad (5.30)$$

from (5.27) and (5.28), we may show that

$$\overline{C} = M_1 B^{-1} \overline{W} \qquad . \qquad (5.31)$$

Furthermore, after some algebra, we find that

$$\underset{\sim}{\Phi} = M^{-1}\underset{\sim}{b} = B^{-1}\left[\underset{\sim}{w} + \overline{W}\ \overline{C}^{-1}(\underset{\sim}{b}_1 - M_1\overline{\underset{\sim}{\Phi}}\ )\right], \qquad (5.32)$$

where

$$B\ \overline{\underset{\sim}{\Phi}} = \underset{\sim}{w} \qquad (5.33)$$

and thus $\underset{\sim}{\Phi}$ may be obtained from the following algorithm.

(i)   Compute $\overline{C} = M_1 B^{-1}\overline{W}$

(ii) Solve $B\ \overline{\underset{\sim}{\Phi}} = \underset{\sim}{w}$ for $\overline{\underset{\sim}{\Phi}}$

(iii) Solve $\overline{C}\ \underset{\sim}{y} = \underset{\sim}{b}_1 - M_1\overline{\underset{\sim}{\Phi}}$ for $\underset{\sim}{y}$ (5.34)

(iv) Solve $B\ \underset{\sim}{\Phi} = \underset{\sim}{w} + \overline{W}\ \underset{\sim}{y}$ for $\underset{\sim}{\Phi}$.

Clearly, if (5.22) may be solved rapidly then steps (ii) and (iv) are similarly efficient. In step (i) the capacity matrix $\overline{C}$ involves the computation of $B^{-1}\overline{W}$ which is just the first p columns of B^{-1} and these may be determined by solving p systems of the form (5.22) with appropriate right-hand sides. We note that the role of $\overline{C}$ in this analysis is essentially that of C^{-1} in Section 5.2. The multiplication by M_1 requires a further $O(p^2n)$ operations, although this will be a relatively small contribution if, as is frequently the case, M_1 is sparse. As indicated earlier the usual approach is to compute $\overline{C}$ and store its LU decomposition as a preprocessing stage. Consequently, if p is not too large, step (iii) is relatively tolerable and the overall solution of the problem is obtained in approximately the time of two solutions of a system of form (5.22). The solution values at points such as x in Figure 5.3

are, of course, unwanted 'extra' data. As noted in Section 5.2, the general approach is well suited to solving a set of problems with different right-hand sides.

Example 5.4.1

As a particularly simple numerical example of the application of the above algorithm we solve Laplace's equation

$$\nabla^2 \phi = 0, \qquad (5.35)$$

over the semi-circular region Γ of Figure 5.4 with Dirichlet boundary conditions as shown. A coarse grid of side h is used for illustrative purposes such that the radius of the semi-circle is 2h and the region is imbedded in a rectangle Γ' with sides 4h and 2h.

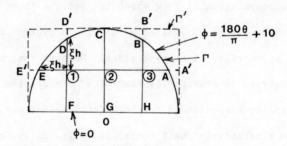

Figure 5.4. The solution domain. The point 0 is the origin of coordinates and θ is measured in the usual anticlockwise sense. At the points A, B, C, D and E the values of ϕ are respectively 40, 70, 100, 130 and 160.

For the grid as shown there are simply three grid points in both R and R', that is, there are no points such as x in Figure 5.3 and thus the equations corresponding to (5.23) may be formulated as follows. At point 2 of the grid we use the usual

5-point approximation and at point 1 we utilise the formula (see, for example, G D Smith p216)

$$\frac{2\Phi_E}{\xi(1+\xi)} + \frac{2\Phi_D}{\xi(1+\xi)} + \frac{2\Phi_2}{1+\xi} + \frac{2\Phi_F}{1+\xi} - \frac{4\Phi_1}{\xi} = 0 \qquad (5.36)$$

where $\xi = \sqrt{3}-1$, and a similar equation at point 3. Thus, for this problem the full system of equations (5.23) may be written as

$$
\begin{aligned}
-\frac{4}{\sqrt{3}-1}\,\Phi_1 + \frac{2}{\sqrt{3}}\,\Phi_2 \qquad\qquad &= -\frac{580}{\sqrt{3}(\sqrt{3}-1)} \\
\Phi_1 - 4\Phi_2 + \Phi_3 \qquad &= -100 \qquad\qquad (5.37) \\
\frac{2}{\sqrt{3}}\,\Phi_2 - \frac{4}{\sqrt{3}-1}\,\Phi_3 &= -\frac{220}{\sqrt{3}(\sqrt{3}-1)} \quad ,
\end{aligned}
$$

or

$$
\begin{bmatrix}
-\dfrac{4}{\sqrt{3}-1} & \dfrac{2}{\sqrt{3}} & 0 \\[2mm]
1 & -4 & 1 \\[2mm]
0 & \dfrac{2}{\sqrt{3}} & -\dfrac{4}{\sqrt{3}-1}
\end{bmatrix}
\begin{bmatrix}
\Phi_1 \\[2mm] \Phi_2 \\[2mm] \Phi_3
\end{bmatrix}
= -
\begin{bmatrix}
457.43 \\[2mm] 100 \\[2mm] 173.51
\end{bmatrix} , \qquad (5.38)
$$

where the right-hand side is correct to 2 decimal places. The purpose of this example is to show how the solution of (5.38) may be obtained using (5.34) and this we now proceed to do.

We extend the boundary data to Γ' by arbitrarily setting $\overline{\Phi} = 100$ at the points A', B', D', and E'. There is no need to extend the definition of the right-hand side of (5.35), although a single halving of the grid would make this necessary. The equations representing (5.35) in the region with boundary Γ' [that is, equations (5.33)] are easily shown to be

$$\begin{bmatrix} -4 & 1 & 0 \\ 1 & -4 & 1 \\ 0 & 1 & -4 \end{bmatrix} \begin{bmatrix} \overline{\Phi}_1 \\ \overline{\Phi}_2 \\ \overline{\Phi}_3 \end{bmatrix} = - \begin{bmatrix} 200 \\ 100 \\ 200 \end{bmatrix} , \tag{5.39}$$

so that $\quad \overline{\Phi}_1 = \overline{\Phi}_3 = 64.29$, $\overline{\Phi}_2 = 57.14.$ $\qquad$ (5.40)

Clearly the first and last rows of the coefficient matrices M and B of systems (5.38) and (5.39) differ and thus the matrix $B^{-1}\overline{W}$ in $\left(5.34(i)\right)$ is a 3x2 matrix whose columns are the first and last columns of B^{-1}, so that

$$\overline{W} = \begin{bmatrix} 1 & 0 \\ 0 & 0 \\ 0 & 1 \end{bmatrix} .$$

Hence we find that

$$B^{-1}\overline{W} = - \frac{1}{56} \begin{bmatrix} 15 & 1 \\ 4 & 4 \\ 1 & 15 \end{bmatrix} , \tag{5.41}$$

and since

$$M_1 = \begin{bmatrix} \dfrac{-4}{\sqrt{3}-1} & \dfrac{2}{\sqrt{3}} & 0 \\ 0 & \dfrac{2}{\sqrt{3}} & \dfrac{-4}{\sqrt{3}-1} \end{bmatrix} , \tag{5.42}$$

we obtain

$$\overline{C} = \frac{1}{14(3-\sqrt{3})} \begin{bmatrix} 13\sqrt{3}+2 & 2-\sqrt{3} \\ 2-\sqrt{3} & 13\sqrt{3}+2 \end{bmatrix} . \tag{5.43}$$

Thus from $(5.34(\text{iii}))$

$$\overline{C}\,\underline{y} = \begin{bmatrix} -457.43 \\ -173.51 \end{bmatrix} - \begin{bmatrix} -\dfrac{4}{\sqrt{3}-1} & \dfrac{2}{\sqrt{3}} & 0 \\ 0 & \dfrac{2}{\sqrt{3}} & -\dfrac{4}{\sqrt{3}-1} \end{bmatrix} \begin{bmatrix} 64.29 \\ 57.14 \\ 64.29 \end{bmatrix}, \quad (5.44)$$

so that $\qquad\qquad y_1 = -125.55, \; y_2 = 82.30,$ $\qquad\qquad (5.45)$

and, from step (iv)

$$\underline{\Phi} = \begin{bmatrix} \Phi_1 \\ \Phi_2 \\ \Phi_3 \end{bmatrix} = \begin{bmatrix} 64.29 \\ 57.14 \\ 64.29 \end{bmatrix} - \frac{1}{56} \begin{bmatrix} 15 & 1 \\ 4 & 4 \\ 1 & 15 \end{bmatrix} \begin{bmatrix} -125.55 \\ 82.30 \end{bmatrix},$$

$$= \begin{bmatrix} 96.45 \\ 60.23 \\ 44.49 \end{bmatrix}. \qquad\qquad (5.46)$$

It may easily be verified that (5.46) is the solution of system (5.38). In step (iv) we made use of (5.40) and (5.41) which, in general, necessitates the storage of $\overline{\underline{\Phi}}$ and $B^{-1}\overline{W}$. In practice the more usual procedure is to re-solve $(5.34(\text{iv}))$ since these equations may be solved quickly and the matrix $B^{-1}\overline{W}$ may be full.

5.5. Applications using splitting rather than imbedding

The imbedding technique described in the preceeding Sections can be very wasteful for certain types of problem. For example, the region R may be such that imbedding it in a rectangle results in a large number of extra unwanted unknowns, such as those at the points x in Figure 5.3. Furthermore, if the differential operator or mesh size changes in various parts of the region, an appropriate splitting of the problem can be advantageous.

As an illustration of the first of these we consider a region R consisting of two rectangles $R^{(1)}$ and $R^{(2)}$, as shown in Figure 5.5, in which we wish to solve (5.1). Here we write this equation in the form

$$L_h \Phi = \zeta \quad \text{in } R_h ,\tag{5.47}$$

where R_h denotes the discrete interior of the grid, L_h denotes a suitable finite-difference replacement for L and we assume that

$$\Phi = g \quad \text{on } \Gamma_h ,\tag{5.48}$$

where Γ_h denotes the discrete boundary of the grid.

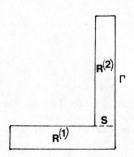

Figure 5.5. **Region of solution of (5.47).**

Clearly from (5.47) and (5.48) we may produce a complete system of linear equations for the nodal Φ values in R_h, in the usual fashion. However, rather than solving these equations, to solve the problem more efficiently, assuming that we can easily solve (5.1) in a rectangle, we can define the matrix B to be the same as the coefficient matrix for the full problem except that on S_h we replace the equations (5.47) by the equations $\overline{\Phi} = g$, where g has been arbitrarily extended to S_h.

Thus the system $B \bar{\underset{\sim}{\Phi}} = \underset{\sim}{w}$ represents the two rectangular problems

$$\left.\begin{array}{l} L_h \, \bar{\Phi}^{(i)} = \zeta \text{ in } R_h^{(i)} \\[2ex] \bar{\Phi}^{(i)} = g \text{ on } \Gamma_h^{(i)} \end{array}\right\} \qquad (i = 1,2) \qquad (5.49)$$

where $\Gamma_h^{(i)}$ denotes the boundary of $R_h^{(i)}$, and each of (5.49) may be easily solved. The capacitance matrix method can then be used to solve the original problem.

As a second example we consider solving the equation

$$\frac{\partial}{\partial x}\left[\alpha(x)\,\frac{\partial\phi}{\partial x}\right] + \frac{\partial}{\partial y}\left[\beta(y)\,\frac{\partial\phi}{\partial y}\right] = \zeta(x,y), \qquad (5.50)$$

over the rectangular region R shown in Figure 5.6
where

$$\alpha(x) = \begin{bmatrix} \alpha_1(x) & \text{in } R^{(1)} \\ \alpha_2(x) & \text{in } R^{(2)} \end{bmatrix}. \qquad (5.51)$$

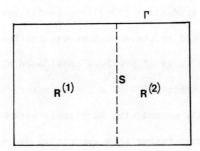

Figure 5.6. The region in which (5.50) is solved.

We assume that $\phi = g$ on Γ and that the functions $\alpha_1(x)$, $\alpha_2(x)$ and $\beta(y)$ are strictly positive and also well-behaved. Furthermore, we suppose that we require $\alpha(\partial\phi/\partial x)$ to be continuous across the boundary S between $R^{(1)}$ and $R^{(2)}$. The overall procedure is very similar to that of the previous example except that the difference equations expressing the continuity of $\alpha(\partial\phi/\partial x)$ across S_h are replaced by $\overline{\Phi} = g$ on S_h. Thus the system $B\,\underset{\sim}{\overline{\Phi}} = \underset{\sim}{w}$ corresponds to the two rectangular problems

$$\frac{\partial}{\partial x}\left[\alpha_i(x)\,\frac{\partial\overline{\phi}^{(i)}}{\partial x}\right] + \frac{\partial}{\partial y}\left[\beta(y)\,\frac{\partial\overline{\phi}^{(i)}}{\partial y}\right] = \zeta(x,y) \text{ in } R^{(i)} \left.\vphantom{\begin{array}{c}1\\1\\1\end{array}}\right\} \quad (5.52)$$
$$\overline{\phi}^{(i)} = g \quad \text{ on } \Gamma^{(i)} ,$$

where $i = 1,2$.

5.6. Discussion

Buzbee et al (1971) give some comparative results using the method of Section 5.4, for solving Poisson's equation over a region consisting of the unit square with a symmetrically placed inner square removed from its centre. Two sizes of inner square were considered and two grid sizes (32x32 and 64x64) were used. The direct solver employed by these authors was the Bunemann algorithm, variant 1. For the range of problems considered their results show that the method is typically 5 to 8 times faster than the Peaceman-Rachford ADI method with Wachspress parameters and roughly 20 to 40 times faster than successive over-relaxation where the greater increase in speed was achieved for the smaller (64x64) grid, for both problems. These comparisons do not include the preprocessing times which are, of course, relatively large. The data given by Buzbee et al indicate that the ratio of preprocessing time to execution time varied from about 6.5 for the 32x32 grid with p = 16 (smaller inner square) to approximately 26 for the 64x64 grid with p = 64 (larger inner square).

Buzbee and Dorr (1974) use a generally similar approach to solving the biharmonic equation on a rectangular region (rectangular clamped plate problems) and use the method of matrix decomposition as a direct solver. These authors take advantage of the sparsity of the data when generating the capacity matrix and also utilise the fact that the solutions are required at only p points. Furthermore, by applying these ideas to the Poisson problem considered previously (with the Bunemann algorithm as fast solver) they show that an overall gain in speed of about 3 is attainable.

By considering some aspects of potential theory Proskurowski and Widlund (1976) suggested that iterative methods, such as the conjugate gradient method, might be well-suited to solving the capacitance matrix equations. This was found to be borne out in practice, particularly for larger problems or if only one or two problems need to be solved. These workers provide a detailed discussion of the relative merits of the methods used by various workers [see also Proskurowski (1980)] and O'Leary and Widlund (1979) extend these ideas to three-dimensional problems.

Techniques which avoid the explicit generation and storage of the capacitance matrix (at some cost in terms of overall computational effort) have been described by Proskurowski (1979) and capacitance matrix methods have been used in finite-element calculations by, for example, Dryja (1982) and Proskurowski and Widlund (1980), and also in certain eigenvalue problems by Proskurowski (1978).

CHAPTER 6
Two Methods for More General Problems

6.1. Introduction

We have seen how FFT techniques and their sophisticated developments enable us to solve various partial differential equations speedily and conveniently on a digital computer. The advent of parallel processors has rendered this a fruitful area of applied research. As future developments seem to be without limit we conclude the present volume by describing just two applications of the material of earlier Chapters. In particular we discuss the use of fast-solvers for the solution of non-separable elliptic equations and the solution of linear time-dependent problems involving two spatial variables and time, using an efficient Laplace transform technique. Throughout this Chapter we shall, for simplicity, assume that the spatial domain is rectangular although this restriction can, at least in principle, be removed by utilising the techniques discussed in Chapter 5 together with the methods described in the following Sections.

There are, of course, many other topics which could be included here. For example, marching methods, as developed for Poisson's equation by Lorenz (1976) and the general separable

equation by Bank (1977a,b) are competitive fast-solvers. The use
of fast-solvers in the solution of the N-body problem has been
discussed, for example by Hockney et al (1974), Eastwood (1975), and
methods for isolated systems are considered by James (1977).
Furthermore, pseudo-spectral methods and multigrid methods can
provide efficient methods of solution for more general problems and
the review articles by Gottlieb and Orsag (1977) and Brandt (1984)
provide an introduction to these topics. A good general guide to
recent developments is given by Birkhoff and Schoenstadt (1984).

6.2. Nonseparable elliptic problems

We first consider an iterative procedure formulated by Concus
and Golub (1973) for strongly elliptic equations of divergence form
on a rectangular region R. Their technique is based on a modified
form of the iteration

$$-\nabla^2 \phi_{n+1} = -\nabla^2 \phi_n - \tau(L\phi_n - \zeta) , \qquad (6.1)$$

for solving

$$L\phi = \zeta , \qquad (6.2)$$

where τ is a parameter and n the iteration counter. The form of
iteration (6.1) is that considered by D'Yakonov (1961) and also
described by Gunn (1965) and Widlund (1972). For $\tau=1$ (6.1) is
equivalent to the 'intuitive' iteration obtained by writing
$L\phi = -\nabla^2\phi - \overline{L}\phi$ and solving $-\nabla^2\phi_{n+1} = \overline{L}\phi_n + \zeta$.

In particular Concus et al consider in detail equation (6.2)
with

$$L\phi \equiv -\underset{\sim}{\nabla} . \left[D(x,y)\underset{\sim}{\nabla}\phi \right] , \qquad (6.3)$$

where the isotropic "diffusion coefficient" $D(x,y)$ is strictly positive on R and its boundary Γ and

$$\phi = g \quad \text{on} \ \Gamma \quad . \tag{6.4}$$

The procedure (6.1) can be too slowly convergent for practical purposes and by scaling the original problem and using a shifted form of (6.1) on the scaled problem Concus et al show that a considerable improvement in convergence rate can be achieved. Additionally the application of Chebyshev acceleration can further improve the convergence rate in some examples.

In the following discussion we assume smooth $D(x,y)$, although the technique can be applied to problems with piecewise smooth $D(x,y)$ as outlined later. We set

$$w(x,y) = \left[D(x,y) \right]^{1/2} \phi(x,y) \ , \tag{6.5}$$

and assuming that $D^{1/2}$ is twice differentiable we obtain from (6.2) and (6.3)

$$Mw \equiv -\nabla^2 w + p(x,y)w = q(x,y) \ , \tag{6.6}$$

where

$$\left. \begin{array}{l} p(x,y) = D^{-1/2} \nabla^2 (D^{1/2}) \\ q(x,y) = D^{-1/2} \zeta(x,y) \end{array} \right] . \tag{6.7}$$

Thus relation (6.5) transforms the operator L into one with differential part $-\nabla^2$ and we note that in obtaining (6.6) we have divided both sides by $D^{1/2}$. The shifted form of (6.1) which may be used to solve $Mw = q$ may be written as

$$(-\nabla^2+K)w_{n+1} = (-\nabla^2+K)w_n - \tau(Mw_n-q), \qquad (6.8)$$

where K is a constant and using (6.6) we obtain the iteration

$$(-\nabla^2+K)w_{n+1} = (-\nabla^2+K)w_n - \tau(-\nabla^2+p)w_n + \tau q . \qquad (6.9)$$

Concus et al show that the choice

$$K = \frac{1}{2}(\min p + \max p) , \qquad (6.10)$$

leads to the optimal estimate

$$\tau = 1 , \qquad (6.11)$$

for the iteration parameter, where min p and max p denote respectively the minimum and maximum values of $p(x,y)$ over the closed rectangle. Thus relation (6.9) simplifies to the form

$$(-\nabla^2+K)w_{n+1} = (K-p)w_n + q . \qquad (6.12)$$

A discrete form of (6.12) may be obtained in the standard fashion using the usual 5-point approximation to the Laplacian together with the boundary condition

$$w = D^{1/2}g \text{ on } \Gamma , \qquad (6.13)$$

and hence a fast-direct method of solution may be used in each iteration. For smooth $D(x,y)$ Concus et al show that the rate of convergence of iteration (6.12) is essentially independent of mesh size.

Chebyshev acceleration (see, for example, Varga, Chapter 5) may be applied to the discrete analogue of (6.12) in the form

$$\tilde{\underset{\sim}{W}}^{(n+1)} = \omega_{n+1}\left(\underset{\sim}{W}^{(n+1)} - \tilde{\underset{\sim}{W}}^{(n-1)}\right) + \tilde{\underset{\sim}{W}}^{(n-1)}, \qquad (6.14)$$

where $\underset{\sim}{W}^{(n+1)}$ denotes the vector of nodal w-values at the (n+1)th stage, $\tilde{\underset{\sim}{W}}^{(n+1)}$ is the improved value of $\underset{\sim}{W}^{(n+1)}$ and $\underset{\sim}{W}^{(n+1)}$ satisfies the discrete form of (6.12) with $\tilde{\underset{\sim}{W}}^{(n)}$ on the right-hand side. We choose $\tilde{\underset{\sim}{W}}^{(-1)} = \underset{\sim}{0}$, $\tilde{\underset{\sim}{W}}^{(0)} = \underset{\sim}{W}^{(0)}$ and set

$$\left.\begin{array}{ll} \omega_1 = 1 & , \\[2mm] \omega_2 = \dfrac{2}{2-\rho^2} & , \\[4mm] \omega_{n+1} = \dfrac{1}{(1-\rho^2\omega_n/4)} & , \qquad n = 2,3,\ldots \end{array}\right\} \qquad (6.15)$$

where ρ denotes the spectral radius of the appropriate iteration matrix which, in general, can be estimated as the iteration proceeds from the formula

$$\rho = \left|\left|\underset{\sim}{W}^{(n)} - \underset{\sim}{W}^{(n-1)}\right|\right| / \left|\left|\underset{\sim}{W}^{(n-1)} - \underset{\sim}{W}^{(n-2)}\right|\right|, \qquad (6.16)$$

in the usual fashion. We note here that obtaining precise optimal
Chebyshev parameters is not of crucial importance in the overall
approach since the scaling and shifting alone are often sufficient
to achieve a rapid convergence rate.

Several practical applications are given in Concus et al for
problems in which $w(x,y) = 2\left[(x-1/2)^2 + (y-1/2)^2\right]$ and with various
choices of $D(x,y)$. For example, for the smooth case
$D(x,y) = \left[1 + (x^4+y^4)/2\right]^2$, for which the maximum and minimum values
of $p(x,y)$ are respectively 6 and 0 so that, from (6.10) K=3, the
maximum error after 5 iterations is shown in Table 6.1 for the
method with and without Chebyshev acceleration.

K	Chebyshev acceleration	Maximum error
3	none	3.9,-8
3	using(6.14)-(6.16)	4.3,-9

Table 6.1 Maximum errors after 5 iterations for
$D(x,y) = \left[1 + (x^4+y^4)/2\right]^2$.

These calculations were carried out for a 64 x 64 grid (with $\tau=1$)
and initial guess $\underset{\sim}{w}^{(0)} = \underset{\sim}{0}$, which gives rise to an initial maximum
error of approximately unity.

For non-smooth D(x,y) the convergence properties of the
iteration resulting from scaling and shifting are not independent
of the mesh size and furthermore, it is more difficult to determine
a reasonable estimate for K. For example, for the case
$D(x,y) = (1+4|x-1/2|)^2$ which has a discontinuous slope at $x = 1/2$,
Concus et al found by experiment that, for a 16 x 16 grid with
$\tau = 1$, a local minimum for ρ occurred at about K = 13, whereas using
(6.10) in an appropriate sense, that is using only grid-point
values of p, suggests K $\simeq$ 53.5. Their results after 10 iterations
(including Chebȳshev acceleration) are shown in Table 6.2.

K	Maximum error
53.5	2.2,-3
13	1.2,-7

Table 6.2 Maximum errors after 10 iterations for
$D(x,y) = (1+4|x-1/2|)^2$.

Clearly the rate of convergence is highly sensitive to the proper
choice of the parameter K in such problems.

Bank (1977) has suggested a similar iterative approach to anisotropic problems of the form (6.2) with

$$L\phi = -\frac{\partial}{\partial x}\left(D_1(x,y)\frac{\partial \phi}{\partial x}\right) - \frac{\partial}{\partial y}\left(D_2(x,y)\frac{\partial \phi}{\partial y}\right) + \left(q_1(x,y) + q_2(x,y)\right)\phi \text{ in R}$$

(6.17)

and $\phi = g$ on Γ, (6.18)

where $D_1, D_2 > 0$ and $q_1, q_2 \geq 0$ in R. After suitable differencing (6.17) and (6.18) may be expressed in the matrix form

$$A\underset{\sim}{z} = \underset{\sim}{b} ,$$

(6.19)

which Bank solves using the iteration

$$\hat{A}\underset{\sim}{z}^{(n+1)} = \hat{A}\underset{\sim}{z}^{(n)} - \tau(A\underset{\sim}{z}^{(n)} - \underset{\sim}{b}),$$

(6.20)

where $\hat{A}$ is a matrix which can be inverted by a fast-direct solver. The form of iteration (6.20) is generally similar to (6.1) or (6.8) and is a form of D'Yakonov-Gunn iteration.

Bank's choice of $\hat{A}$ is that for a 'nearby' separable equation [see (3.95)] which can be solved rapidly by marching or by Swartztrauber's cyclic reduction algorithm. In principle we may choose $\hat{A}$ as we please; for example, $\hat{A}$ could be chosen to be the matrix arising from the $(-\nabla^2 + K)$ operator as in Concus et al. Such a choice is likely to reduce the computation time per iteration, since the fast methods for the general separable equation are

typically 2 to 5 times slower than, for example, Temperton's
Poisson-solver, but more iterations are likely to be necessary.
The best strategy will depend on the problem being solved.

Scaling can be introduced in order to improve the convergence
rate of (6.20) by setting

$$s(x,y) \ w(x,y) = \phi(x,y), \qquad (6.21)$$

and solving the equation for $w(x,y)$ instead of $\phi(x,y)$. The results
for a number of test problems and different scalings are given by
Bank using the value $\tau = 1$. In all cases considered it was found
possible to choose $s(x,y)$ so that after 10 iterations the number of
correct digits was roughly double that for the unscaled iteration.
We note here that the Concus and Golub iteration is a special case
of Bank's form of D'Yakanov-Gunn iteration with $D_1 = D_2 = D$ and
$s(x,y) = D^{-1/2}$.

6.3. Use of the Laplace transform for linear time-dependent problems

In this Section we show how Laplace transform techniques can
be applied to efficiently determine the numerical solution of a
certain class of linear parabolic and hyperbolic equations
involving as independent variables, time t, and two spatial
variables x and y. The basic idea follows the usual Laplace
transform approach in that after applying the transform to the
differential equation the number of independent variables is

reduced by one and the resulting 'subsidiary' equation involves a complex parameter s instead of the variable t. Solving this equation gives the Laplace transform of the solution as a function of x and y and the solution of the original problem is then obtained by applying the inverse Laplace transform.

In practice, the numerical application of a suitable inversion procedure means that the subsidiary equation has to be solved for several different values of the parameter s, in order to obtain the solution of a parabolic or hyperbolic problem at some specified time $t = \bar{t}$. Thus it is desirable that the subsidiary equation can be rapidly solved and Huntley et al (1978) consider problems in which this equation and boundary conditions are of a form which allows the use of the FFT techniques described in Chapter 3. Hence the spatial domain is taken to be rectangular and, furthermore, it is easily deduced that the most general form of parabolic or hyperbolic equation which, after Laplace transformation, leads to a suitable form of subsidiary equation, may be written as

$$\left[f_1(D)\frac{\partial^2}{\partial x^2} + f_2(D)\frac{\partial^2}{\partial y^2} + f_3(D)\frac{\partial}{\partial x} + f_4(D)\frac{\partial}{\partial y} + f_5(D) \right] \phi(x,y,t) = \zeta(x,y,t), \tag{6.22}$$

where $D \equiv \partial/\partial t$ and the operators $f_i(D)$ $(i=1,2,\ldots,5)$ are polynomial operators of the form

$$f_i(D) = a_{i,r}D^r + a_{i,r-1}D^{r-1} + \ldots + a_{i,0} , \tag{6.23}$$

where the coefficients $a_{i,j}$ may be, at most, functions of y for

i=2,4 or 5.

Denoting by $\psi(x,y,s)$ the Laplace transform of $\phi(x,y,t)$ so

that

$$\psi(x,y,s) = \int_0^\infty \phi \, e^{-st} dt \equiv T\left[\phi\right], \qquad (6.24)$$

it follows that

$$T\left[D^j \frac{\partial^k \phi}{\partial y^k}\right] = s^j \frac{\partial^k \psi}{\partial y^k} - s^{j-1}\frac{\partial^k \phi}{\partial y^k} - s^{j-2}D\frac{\partial^k \phi}{\partial y^k} - \ldots - D^{j-1}\frac{\partial^k \phi}{\partial y^k}, \qquad (6.25)$$

$$(k=0,1,2; \; j=1,2,\ldots,r)$$

where, in the usual manner, terms involving ϕ and its derivatives

in (6.25) have to be specified at t = 0. Thus the transform of a

typical term in (6.22) may be written as

$$T\left[f_2(D) \frac{\partial^2 \phi}{\partial y^2}\right] = f_2(s) \frac{\partial^2 \psi}{\partial y^2} + h_2(x,y,s) \;,$$

where $h_2(x,y,s)$ represents all the initial condition information

and hence the subsidiary equation associated with (6.22) has the

form

$$\frac{\partial^2 \psi}{\partial x^2} + \frac{f_2(s)}{f_1(s)} \frac{\partial^2 \psi}{\partial y^2} + \frac{f_3(s)}{f_1(s)} \frac{\partial \psi}{\partial x} + \frac{f_4(s)}{f_1(s)} \frac{\partial \psi}{\partial y} + \frac{f_5(s)}{f_1(s)} \psi$$

$$= \frac{1}{f_1(s)} \left[\eta(x,y,s) + \sum_{i=1}^{4} h_i(x,y,s) \right] , \qquad (6.26)$$

where $\eta = T[\zeta]$. It may be shown that it is reasonable to assume that $f_1(s) \neq 0$ for the values of s employed in the inversion procedure and, in particular, Huntley et al consider problems in which $f_2(s)/f_1(s) = 1$ and $f_3 = f_4 \equiv 0$ so that (6.26) has the form of a Helmholtz equation

$$\nabla^2 \psi + S \psi = q , \qquad (6.27)$$

where ψ, S and q are complex. That (6.26) could be solved, in general, by the methods of Chapter 3 should be clear by comparing this equation with equation (3.10). Such general problems were not tackled by Huntley et al and the overall procedure for such cases is likely to be complicated particularly by the possibility that $\text{sgn}[\text{Re}(f_2/f_1)]$ may vary with s.

We suppose that Dirichlet boundary conditions are imposed, so that

$$\phi(x,y,t) = g(x,y,t) , \qquad (6.28)$$

on the boundary of the spatial domain and hence in solving (6.27)
we use the condition

$$\psi(x,y,s) = G(x,y,s) , \qquad (6.29)$$

where $G = T[g]$. Thus, for a given s, (6.27) may be solved by
FFT methods (using the usual sine-transform) and the only
complication compared with the methods of Chapter 3 is that the
Fourier harmonics are complex. Thus we have to solve complex
tridiagonal systems. Both 5-point and 9-point approximations to
(6.27) were used by Huntley et al and a single cyclic reduction
stage was incorporated into both algorithms.

A number of methods are available for numerical inversion of
the Laplace transform and the results of Huntley et al were
obtained using an efficient algorithm devised by Vlach (1969) and
Singhal and Vlach (1971, 1975, 1976) which is based on the Padé
approximation to e^z. Formally

$$\phi(x,y,t) = \frac{1}{2\pi i} \int_{c-i\infty}^{c+i\infty} \psi(x,y,s) \, e^{st} \, ds , \qquad (6.30)$$

where c is a real constant greater than the real part of any pole
of $\psi(x,y,s)$ and substituting

$$z = st, \qquad (6.31)$$

relation (6.30) takes the form

$$\phi(x,y,t) = \frac{1}{2\pi i t} \int_{c'-i\infty}^{c'+i\infty} \psi(x,y,z/t)e^z dz \quad . \quad (6.32)$$

By approximating e^z in the form

$$e^z \simeq R_{N,M}(z) = P_N(z)/Q_M(z) \quad , \quad (6.33)$$

where $P_N(z)$ and $Q_M(z)$ are respectively polynomials of degree N and M, Vlach (1969) showed that all the poles z_j of $R_{N,M}(z)$ are simple poles and hence, using partial fractions,

$$R_{N,M}(z) = \sum_{j=1}^{M} K_j/(z-z_j) \quad , \quad (6.34)$$

where K_j denotes the relevant residue of $R_{N,M}(z)$. Thus, completing the contour by a semi-circular arc at infinity, it follows from (6.32)-(6.34) that

$$\hat{\phi}(x,y,t) = -\frac{1}{t} \sum_{j=1}^{M} K_j \psi(x,y,z_j/t) \quad , \quad (6.35)$$

where $\hat{\phi}$ denotes an approximation to ϕ consequent upon the use of (6.33) and it is assumed that the contribution to (6.32) from the semi-circular arc is zero. Singhal et al (1975) have shown that this is the case for the choice $M \geqslant N + 2$ and values of K_j and z_j are tabulated by Singhal et al (1976) for various M and N. For even values of M all the roots of $Q_M(z)$ occur as complex conjugate pairs and (6.35) may be written as

$$\hat{\phi}(x,y,t) = -\frac{2}{t} \sum_{j=1}^{M'} \text{Re}\left[K_j\psi(x,y,z_j/t)\right] , \qquad (6.36)$$

where M' = M/2. It may be shown that (6.36) [or (6.35)] inverts exactly the first M + N + 1 terms of the appropriate Taylor expansion of $\phi(x,y,t)$ and it is clearly computationally advantageous to use (6.36) since, for a given choice of M, only half the number of subsidiary equation solutions need be obtained compared with (6.35).

Thus to obtain the solution of an equation of the general form (6.22) at $t = \bar{t}$, the overall procedure, after Laplace transformation and suitable choice of M', may be summarised as:

For j = 1,2,...,M'

(i)   calculate $s_j = z_j / \bar{t}$

(ii)   solve the subsidiary equation (6.26) for $\psi(x,y,s_j)$ with boundary condition (6.29)

(iii) accumulate the final solution via (6.36).

As indicated earlier, Huntley et al concentrated on problems for which the subsidiary equation has the particular form (6.27) and for these problems the values M'=2 and 5 (with N=M-2) were found to be adequate. In fact as M' increases the residues K_j grow rapidly in magnitude and loss of significance in K_j due to computer round-off error can affect the results. This was found to occur

for calculations performed with M' = 8 on an ICL 1906S machine
which has a precision of approximately 11 decimal digits, the
results in such cases being slightly inferior to those for M' = 5.
In principle, increasing M' essentially increases the size of the
step in time which is feasible but, in practice, the largest value
of M' which can be used is limited by the above considerations.
Since it is a straightforward matter to evaluate time-derivatives of
the solution using the Laplace transform technique, for example,

$$\frac{\partial \hat{\phi}}{\partial t}(x,y,t) = - \frac{1}{t} \sum_{j=1}^{M} \frac{z_j}{t} K_j \psi(x,y,z_j/t) , \qquad (6.37)$$

a sensible practical approach is to choose say M' = 2, evaluate the
solution and its necessary time-derivatives at $t = \bar{t}$ and restart
the algorithm from this point with further steps in a similar
manner. The best strategy will vary from problem to problem and
machine to machine, but usually in such problems a time-history of
the solution is required.

As indicated earlier, the method can handle both parabolic and
hyperbolic equations with only very minor programming changes. For
example it is trivially shown that the equations

$$\frac{\partial^2 \phi}{\partial x^2} + \frac{\partial^2 \phi}{\partial y^2} = \frac{\partial^2 \phi}{\partial t^2} , \qquad (6.38)$$

and
$$\frac{\partial^2 \phi}{\partial x^2} + \frac{\partial^2 \phi}{\partial y^2} = \frac{\partial \phi}{\partial t} , \qquad (6.39)$$

both reduce to the form (6.27) after Laplace transformation and practical details for such problems (with boundary conditions independent of t) are given by Huntley et al.

Problems with time-varying source terms and boundary conditions were also considered by these authors and we note that such problems are no more difficult to handle by this method than those with time invariant boundary conditions and source terms. Huntley et al provide detailed results for various grid sizes and different values of M' for the equation

$$\frac{\partial \phi}{\partial t} = \frac{\partial^2 \phi}{\partial x^2} + \frac{\partial^2 \phi}{\partial y^2} + \zeta(x,y,t), \qquad (6.40)$$

which is satisfied over the square $0 \leqslant x \leqslant 2$, $0 \leqslant y \leqslant 2$ with

$$\zeta(x,y,t) = \sin x \exp(-t)/(1+y)^2 - 2x - 6xy, \qquad (6.41)$$

and initial and time-varying boundary conditions derived from the analytic solution

$$\phi(x,y,t) = \sin x \ln(1+y) \exp(-t) + x^3 y + xy^2 . \qquad (6.42)$$

Three grid sizes were used and the results using the five-point FFT method show that for $0 < \overline{t} \leqslant 2$, the mid-point errors reduce by a factor of about 4 for each halving of the spatial grid and, furthermore, that the choice M' = 2 or M' = 5 gives very similar error magnitudes for this range of $\overline{t}$. This leads to the conclusion that, for this problem with M' = 2, the errors due to the Laplace transform inversion become predominant for $\overline{t} > 2$ since reasonably satisfactory results were obtained with M' = 5 for $0 < \overline{t} \leqslant 20$. All

the computations were carried out using a single step in time from
t = 0 to the chosen value t = $\bar{t}$ and the above conclusions are
supported by the fact that very similar calculated errors were
obtained for $0 < \bar{t} < 2$ using the Peacemann-Rachford alternating-
direction method with time step $\delta t = 0.01$.

For such problems the execution times given by Huntley et al
show that the LT-FFT method can be very competitive. For the above
example, to compute the solution at t = 1 using the ADI method with
$\delta t = 0.01$ took approximately fourteen times as long as using the
LT-FFT method (with M' = 2) and this factor was found to be roughly
independent of spatial grid size. However, for problems with
time independent boundary conditions the ADI methods were found to
be relatively more efficient.

It is clear that, for suitable problems, the LT-FFT method
works well and, in principle, a much wider class of problems than
those considered by Huntley et al can be solved by this type of
approach. For example, in addition to those problems which lead to
the general two-dimensional form (6.26) of subsidiary equation,
certain parabolic and hyperbolic equations involving three spatial
variables and time could be tackled by this method. Furthermore,
it is possible that more general time-dependent problems could be
solved efficiently by a suitable combination of the LT-FFT method
and the D'Yakonov-Gunn type of iteration outlined in the previous
Section.

Appendix 1

A1.1   Introduction

We consider an N-th order real symmetric matrix M and a real diagonal matrix D with positive diagonal elements, such that

$$M\underset{\sim}{x}_r = \lambda_r D\underset{\sim}{x}_r ,$$ (A1.1)

where λ_r denotes the r-th eigenvalue of $D^{-1}M$ and $\underset{\sim}{x}_r$ denotes the corresponding eigenvector.   It can be proved that all the eigenvalues of $D^{-1}M$ are real and that to each eigenvalue λ_r there corresponds an eigenvector $\underset{\sim}{x}_r$ which can be chosen so that

$$\underset{\sim}{x}_r^T D\underset{\sim}{x}_s = 0, \quad r \neq s$$ (A1.2)

that is, the eigenvectors may be chosen so as to be mutually orthogonal relative to the matrix D.   It is always possible to choose such a set of eigenvectors whether or not the eigenvalues are distinct (see, for example, Hildebrand (1968), Section 1.11), and in the following Sections formulae for the eigenvalues and eigenvectors for a number of specialised matrices M and D are derived.

153

A1.2   The Dirichlet problem

For this case the relevant matrix M of order N=n-1 has the symmetric tridiagonal form

$$M = \begin{bmatrix} -2 & 1 & & & \\ 1 & -2 & 1 & & \\ & & \cdots\cdots\cdots & & \\ & & 1 & -2 & 1 \\ & & & 1 & -2 \end{bmatrix} \qquad (A1.3)$$

and D is the unit matrix of order n-1.   The actual matrix which is generated by a particular Dirichlet problem may be some simple function of M, such as the matrix A given by (1.8), which has the same eigenvectors as M but whose eigenvalues are all less than those of M by an amount $2h^2$.

In order to determine the eigenvalues and corresponding eigenvectors of M, we consider the problem

$$\delta^2 \phi_{j,r} - \lambda_r \phi_{j,r} = 0, \quad (j=1,2,\ldots,n-1) \qquad (A1.4)$$

given

$$\phi_{0,r} = \phi_{n,r} = 0, \qquad (A1.5)$$

where $\phi_{j,r}$ denotes the j-th component of the r-th eigenvector $\underset{\sim}{x}_r$, r=1,2,...,n-1, and δ denotes the central-difference operator. Equation (A1.4) may be written in the form

$$\phi_{j-1,r} - 2\cos\theta_r \phi_{j,r} + \phi_{j+1,r} = 0, \qquad (A1.6)$$

where

$$\lambda_r = -2 + 2\cos\theta_r \qquad (A1.7)$$

and the general solution of (A1.6) is easily shown to be

$$\phi_{j,r} = c_1 \cos j\theta_r + c_2 \sin j\theta_r, \qquad (A1.8)$$

where c_1 and c_2 are arbitrary constants.   The conditions (A1.5) require that $c_1 = 0$ and $\theta_r = \pi r/n$, so that

$$\lambda_r = -4\sin^2 \frac{\pi r}{2n}$$

and $\qquad \underset{\sim}{x}_r = \left[\sin\frac{\pi r}{n}, \sin\frac{2\pi r}{n}, \ldots, \sin\frac{(n-1)\pi r}{n} \right]^T.$ \hfill (A1.9)

It is easily verified that

$$\underset{\sim}{x}_r^T \underset{\sim}{x}_s = \frac{1}{2} n \delta_{rs}, \tag{A1.10}$$

where δ_{rs} denotes the Kronecker delta function, and hence the results of this Section are in accordance with the methods described in Chapter 1 [equations (1.11) to (1.18)] and Chapter 2 [equations (2.69) and (2.71)], since if we expand any (n-1) component vector $\underset{\sim}{v}$ in terms of the vectors $\underset{\sim}{x}_r$ so that

$$\underset{\sim}{v} = \sum_{r=1}^{n-1} b(r) \underset{\sim}{x}_r, \tag{A1.11}$$

then $\qquad b(r) = \frac{2}{n} \underset{\sim}{x}_r^T \underset{\sim}{v}, \qquad (r=1,2,\ldots,n-1).$ \hfill (A1.12)

A1.3 The Neumann problem

The relevant matrix $\bar{M}$ of order $N=n+1$ is given by

$$\bar{M} = \begin{bmatrix} -2 & 2 & & & & \\ 1 & -2 & 1 & & & \\ & & \cdots\cdots\cdots & & & \\ & & & 1 & -2 & 1 \\ & & & & 2 & -2 \end{bmatrix} \tag{A1.13}$$

and is related to an appropriate (n+1)th order symmetric tridiagonal matrix M by

$$\bar{M} = D^{-1}M, \tag{A1.14}$$

where

$$
M = \begin{bmatrix}
-1 & 1 & & & & \\
1 & -2 & 1 & & & \\
& & \cdots\cdots\cdots & & & \\
& & & 1 & -2 & 1 \\
& & & & 1 & -1
\end{bmatrix}
$$

and

$$
D = \begin{bmatrix}
1/2 & & & & \\
& 1 & & & \\
& & \cdot & & \\
& & & 1 & \\
& & & & 1/2
\end{bmatrix} \qquad (A1.15)
$$

The eigenvalues and the components of the corresponding eigenvectors of $\bar{M}$ may be derived by considering the problem

$$
\delta^2 \phi_{j,r} - \lambda_r \phi_{j,r} = 0 \qquad (j=0,1,\ldots,n) \qquad (A1.16)
$$

given

$$
\phi_{-1,r} = \phi_{1,r} \qquad \phi_{n+1,r} = \phi_{n-1,r}, \qquad (A1.17)
$$

for $r=0,1,\ldots,n$. Equation (A1.16) has a general solution given by (A1.8), and the conditions (A1.17) imply that $c_2=0$ and $\theta_r=\pi r/n$, giving

$$
\lambda_r = -4\sin^2 \frac{\pi r}{2n}
$$

and

$$
\underset{\sim}{x}_r = \left[1, \cos\frac{\pi r}{n}, \cos\frac{2\pi r}{n}, \ldots, \cos\frac{(n-1)\pi r}{n}, (-1)^r \right]^T \qquad (A1.18)
$$

It may be shown that

$$
\underset{\sim}{x}_r^T D \underset{\sim}{x}_s = \begin{cases} n & r=s=0,n \\ \dfrac{n}{2} & r=s\neq 0,n \\ 0 & r\neq s, \end{cases} \qquad (A1.19)
$$

where D is given by (A1.15), and hence if we express any (n+1) component vector $\underset{\sim}{v}$ in the form

$$\underset{\sim}{v} = \sum_{r=0}^{n}{}'' a(r)\underset{\sim}{x}_r, \qquad (A1.20)$$

where $\sum''$ denotes that the first and last terms in the summation are to be halved in accordance with (2.55), then

$$a(r) = \frac{2}{n}\underset{\sim}{x}_r^T D\underset{\sim}{v}, \quad (r=0,1,\ldots,n) \qquad (A1.21)$$

as indicated by (2.57).

A1.4   The periodic problem

The appropriate matrix M of order N=2n is given by

$$M = \begin{bmatrix} -2 & 1 & & & & 1 \\ 1 & -2 & 1 & & & \\ & & \ldots\ldots\ldots & & & \\ & & & 1 & -2 & 1 \\ 1 & & & & 1 & -2 \end{bmatrix} \qquad (A1.22)$$

and D is the unit matrix of order 2n.

We consider the associated difference equation

$$\delta^2\phi_{j,r} - \lambda_r\phi_{j,r} = 0 \quad (j=0,1,\ldots,2n-1) \qquad (A1.23)$$

and the condition

$$\phi_{2n+m,r} = \phi_{m,r} \quad (m=0,\pm1,\pm2,\ldots) \qquad (A1.24)$$

for $r=0,1,\ldots,2n-1$.

Applying (A1.24) to the general solution of (A1.23), we find that both sine and cosine terms are present for a given λ_r if we choose $\theta_r = \pi r/n$. Denoting by $\underset{\sim}{y}_r$ a vector whose components consist of the cosine terms and by $\underset{\sim}{z}_r$ a vector whose components are the sine terms, we may deduce that there are n-1 eigenvalues each of multiplicity two, given by

$$\lambda_r = -4\sin^2\left[\frac{\pi r}{2n}\right] , \quad (r=1,2,\ldots,n-1) \qquad \text{(A1.25)}$$

corresponding to the eigenvectors

$$\underset{\sim}{y}_r = \left[1,\cos\frac{\pi r}{n},\cos\frac{2\pi r}{n},\ldots,\cos\frac{(2n-1)\pi r}{n} \right]^T ,$$

and $\quad \underset{\sim}{z}_r = \left[0,\sin\frac{\pi r}{n},\sin\frac{2\pi r}{n},\ldots,\sin\frac{(2n-1)\pi r}{n} \right]^T \qquad \text{(A1.26)}$

and two eigenvalues of unit multiplicity given by

$$\lambda_0 = 0$$

and $\qquad \lambda_n = -4,$ $\qquad\qquad\qquad\qquad$ (A1.27)

corresponding to, respectively, the eigenvectors

$$\underset{\sim}{y}_0 = (1,1,1,\ldots,1)^T .$$

and $\qquad \underset{\sim}{y}_n = (1,-1,1,\ldots,-1)^T$ $\qquad\qquad$ (A1.28)

It is easy to verify that

$$\underset{\sim}{y}_r^T \underset{\sim}{z}_s = 0, \quad r=0,1,\ldots,n; \quad s=1,2,\ldots,n-1 \qquad \text{(A1.29)}$$

$$\underset{\sim}{y}_r^T \underset{\sim}{y}_s = \left. \begin{array}{ll} 2n & r=s=0,n \\ n & r=s\neq 0,n \\ 0 & r\neq s \end{array} \right] \qquad \text{(A1.30)}$$

and $\qquad \underset{\sim}{z}_r^T \underset{\sim}{z}_s = n\delta_{rs}.$ $\qquad\qquad\qquad$ (A1.31)

Thus the vectors $\underset{\sim}{y}_r$, $r=0,1,\ldots,n$; $\underset{\sim}{z}_r$, $r=1,2,\ldots,n-1$ form a complete set of mutually orthogonal vectors which may be identified with the vectors $\underset{\sim}{x}_r$, $r=0,1,\ldots,2n-1$ of Section A1.1. Hence writing any 2n component vector $\underset{\sim}{v}$ in the form

$$\underset{\sim}{v} = \frac{1}{2} a(0)\underset{\sim}{y}_0 + \sum_{r=1}^{n-1} \left[a(r)\underset{\sim}{y}_r + b(r)\underset{\sim}{z}_r \right] + \frac{1}{2} a(n)\underset{\sim}{y}_n \qquad \text{(A1.32)}$$

in accordance with (2.49), it is clear that

$$a(r) = \frac{1}{n}\underset{\sim}{y}^T_r\underset{\sim}{v} = \frac{1}{n} \sum_{j=0}^{2n-1} v_j \cos\frac{\pi j r}{n} \;,\; (r=0,1,\ldots,n) \qquad (A1.33)$$

and

$$b(r) = \frac{1}{n}\underset{\sim}{z}^T_r\underset{\sim}{v} = \frac{1}{n} \sum_{j=0}^{2n-1} v_j \sin\frac{\pi j r}{n}, \;\; (r=1,2,\ldots,n-1) \qquad (A1.34)$$

where v_j denotes the j-th component of $\underset{\sim}{v}$. Equations (A1.33) and (A1.34) correspond to relations (2.50) of Section 2.8.

A1.5 The symmetric Dirichlet problem (Dirichlet-Neumann conditions)

The relevant matrix $\bar{M}$ of order N=n is given by

$$\bar{M} = \begin{bmatrix} -2 & 1 & & & & \\ 1 & -2 & 1 & & & \\ & & \cdots\cdots\cdots & & \\ & & & 1 & -2 & 1 \\ & & & & 2 & -2 \end{bmatrix} \qquad (A1.35)$$

and is related to an n-th order symmetric tridiagonal matrix M via a relation of the form of (A1.14) with

$$M = \begin{bmatrix} -2 & 1 & & & & \\ 1 & -2 & 1 & & & \\ & & \cdots\cdots\cdots & & \\ & & & 1 & -2 & 1 \\ & & & & 1 & -1 \end{bmatrix}$$

$$\text{and} \qquad D= \begin{bmatrix} 1 & & & & \\ & 1 & & & \\ & & \cdot & & \\ & & & 1 & \\ & & & & 1/2 \end{bmatrix} \qquad (A1.36)$$

By considering the difference equation

$$\delta^2\phi_{j,r} - \lambda_r\phi_{j,n} = 0 \quad (j=1,2,\ldots,n) \qquad (A1.37)$$

and the conditions

$$\phi_{0,r} = 0 \text{ and } \phi_{n+1,r} = \phi_{n-1,r} \qquad (A1.38)$$

for $r=1,2,\ldots,n$, we find that

$$\lambda_r = -4\sin^2\left[\frac{(2r-1)\pi}{4n}\right] \qquad (A1.39)$$

and

$$\underset{\sim}{x}_r = \left[\sin\frac{(2r-1)\pi}{2n}, \sin\frac{2(2r-1)\pi}{2n}, \ldots, \sin\frac{(n-1)(2r-1)\pi}{2n}, (-1)^{r+1}\right] \qquad (A1.40)$$

where

$$\underset{\sim}{x}_r^T D \underset{\sim}{x}_s = \frac{1}{2}n\delta_{rs}. \qquad (A1.41)$$

Hence any n component vector $\underset{\sim}{v}$ may be expressed as

$$\underset{\sim}{v} = \sum_{r=1}^{n} c(r)\underset{\sim}{x}_r , \qquad (A1.42)$$

where

$$c(r) = \frac{2}{n}\underset{\sim}{x}_r^T D\underset{\sim}{v} \qquad (A1.43)$$

and in component form equations (A1.42) and (A1.43) are, respectively,

$$v_j = \sum_{r=1}^{n} c(r)\sin\frac{j\pi(2r-1)}{2n} \quad (j=1,2,\ldots,n) \qquad (A1.44)$$

and
$$c(r) = \frac{2}{n} \sum_{j=1}^{n-1} v_j \sin\frac{j\pi(2r-1)}{2n} + \frac{1}{n}(-1)^{r+1} v_n \qquad (A1.45)$$
$$(r=1,2,\ldots,n) \ .$$

Swartztrauber (1977) has shown how the Fourier series (A1.44) is related to the periodic case of the previous Section. Defining f_j by

$$f_j = (v_j + v_{n-j})\cos\frac{j\pi}{2n} + (v_j - v_{n-j})\sin\frac{j\pi}{2n} \ , \qquad (A1.46)$$

we may show that

$$f_j = \sum_{r=1}^{n} c(r) \left[d(r,j)\cos\frac{j\pi r}{n} + e(r,j)\sin\frac{j\pi r}{n}\right] \ , \qquad (A1.47)$$

where, for r even

$$d(r,j) = -1$$
$$e(r,j) = \ 1 \qquad\qquad (A1.48)$$

and for r odd

$$d(r,j) = \cos\frac{j\pi}{n} - \sin\frac{j\pi}{n}$$
$$e(r,j) = \cos\frac{j\pi}{n} + \sin\frac{j\pi}{n} \qquad\qquad (A1.49)$$

These results allow (A1.47) to be written in the form

$$f_j = c(1) + \sum_{r=1}^{n/2-1} \left[c(2r+1)-c(2r)\right]\cos\frac{2r\pi j}{n} + \left[c(2r+1) + c(2r)\right]\sin\frac{2r\pi j}{n}$$
$$+ (-1)^{j+1}c(n). \qquad (A1.50)$$

Comparing (A1.50) and (2.49) $\left[$or (A1.32)$\right]$, we see that if n is replaced by n/2 in (2.49) and we set

$$a(0) = 2c(1)$$

$$a(r) = c(2r+1)-c(2r) \Big] \quad (r=1,2,\ldots,n/2-1) \quad (A1.51)$$

$$b(r) = c(2r+1)+c(2r) \Big]$$

and $\qquad a(n/2) = -2c(n),$

then the method of Section 2.9 may be used to compute f_j, $j=1,2,\ldots,n$. The values of v_j may be recovered from the relation

$$v_j = \frac{1}{2}(f_j + f_{n-j})\sin\frac{j\pi}{2n} + \frac{1}{2}(f_j - f_{n-j})\cos\frac{j\pi}{2n} . \qquad (A1.52)$$

In order to compute the discrete Fourier transform of v_j it is clear that we must construct f_j via (A1.46), use the method of Section 2.8 to obtain $a(r)$, $r=0,1,\ldots,n/2$; $b(r)$, $r=1,2,\ldots,n/2-1$ and then solve (A1.51) for the required harmonics.

Appendix 2 FFT Fortran Computer Subroutine

```
       SUBROUTINE FFT(XREAL,XIMAG,N,NU)
       DIMENSION XREAL(N),XIMAG(N)
       N2=N/2
       NU1=NU-1
       K=0
       DO 100 L=1,NU
102    DO 101 I=1,N2
       P=IBITR(K/2**NU1,NU)
       ARG=6.283185*P/FLOAT(N)
       C=COS(ARG)
       S=SIN(ARG)
       K1=K+1
       K1N2=K1+N2
       TREAL=XREAL(K1N2)*C+XIMAG(K1N2)*S
       TIMAG=XIMAG(K1N2)*C-XREAL(K1N2)*S
       XREAL(K1N2)=XREAL(K1)-TREAL
       XIMAG(K1N2)=XIMAG(K1)-TIMAG
       XREAL(K1)=XREAL(K1)+TREAL
       XIMAG(K1)=XIMAG(K1)+TIMAG
101    K=K+1
       K=K+N2
       IF(K.LT.N) GO TO 102
       K=0
       NU1=NU1-1
100    N2=N2/2
       DO 103 K=1,N
       I=IBITR(K-1,NU)+1
       IF(I.LE.K) GO TO 103
       TREAL=XREAL(K)
       TIMAG=XIMAG(K)
       XREAL(K)=XREAL(I)
       XIMAG(K)=XIMAG(I)
       XREAL(I)=TREAL
       XIMAG(I)=TIMAG
103    CONTINUE
       RETURN
       END
       FUNCTION IBITR(J,NU)
       J1=J
       IBITR=0
       DO 200 I=1,NU
       J2=J1/2
       IBITR=IBITR*2+(J1-2*J2)
200    J1=J2
       RETURN
       END
```

Figure A2.1 FFT Fortran Computer Subroutine (E.Oran Brigham, THE FAST FOURIER TRANSFORM, ©1974, p164. Reprinted by Permission of Prentice-Hall, Inc, Englewood Cliffs, New Jersey).

References

Ahlberg, J.H., Nilson, E.N., and Walsh, J.L. (1967) Theory of
splines and their applications. Academic Press, New York.

Allen, D.N. de G. (1962) A suggested approach to finite-
difference representations of differential equations, with an
application to determine temperature distributions near a
sliding contact. Q.J. Mech. Appl. Math. 15, 11.

Angel, E. (1968) Discrete invariant imbedding and elliptic boundary
value problems over irregular regions. J. Math. Anal. Appl.
23, 471.

Bank, R.E. (1977) Marching algorithms for elliptic boundary value
problems II: the variable coefficient case. SIAM J. Numer.
Anal. 14, 950.

Bank, R.E., and Rose, D.J. (1977) Marching algorithms for elliptic
boundary value problems I: the constant coefficient case.
SIAM J. Numer. Anal. 14, 792.

Bergland, G.D. (1968) A fast Fourier transform using base eight
iterations. Math. Computation 22, 275.

Bickley, W.G. (1948) Finite difference formulae for the square
lattice. Quart. J. Mech., 1, 35.

Birkhoff, G., and Schoenstadt, A. (1984) editors, Elliptic Problem
Solvers II, Academic Press, New York.

Bois, P. and Vignes, J. (1982) An Algorithm for Automatic Round-off
Error Analysis in Discrete Linear Transforms, Intern. J.
Computer Math. 12, 161.

165

Brandt, A. (1984) Multigrid techniques: 1984 guide with
applications to fluid dynamics. Computational Fluid Dynamics
Lecture Series, Von-Karman Inst. for Fluid Dynamics, Belgium.

Brigham, E.O. (1974) The fast Fourier transform, Prentice-Hall,
Englewood Cliffs, New Jersey.

Bruce, G.H., Peaceman, D.W., Rachford, H.H., and Rice, J.D. (1953)
Trans. Am. Inst. Min. Engrs. (Petrol Div.) 198, 79.

Bunemann, O. (1969) A compact non-iterative Poisson-solver. Report
294, Stanford University Institute for Plasma research,
Stanford, California.

Buzbee, B.L. and Dorr, F.W. (1974) The direct solution of the
biharmonic equation on rectangular regions and the Poisson
equation on irregular regions. SIAM J. Numer. Anal. 11, 753.

Buzbee, B.L., Dorr, F.W., George, J.A. and Golub, G.H. (1971) The
direct solution of the discrete Poisson equation on irregular
regions. SIAM J. Numer. Anal. 8, 722.

Buzbee, B.L., Golub, G.H. and Nielson, C.W. (1970) On direct
methods for solving Poisson's equations. SIAM J. Numer.
Anal. 7, 627.

Challis, N.V. and Burley, D.M. (1982) A numerical method for
conformal mapping, IMA J. Numer. Anal. 2, 169.

Christiansen, J.P. and Hockney, R.W. (1971) Delsqphi, A two-
dimensional Poisson-solver program, Comput. Phys. Commun. 2,
139.

Cochran, W.T. (1967) et al What is the fast Fourier transform?
IEEE Trans. Audio and Electroacoustics, 15, 45.

Concus, P., and Golub, G.H. (1973) Use of fast direct methods for the efficient numerical solution of nonseparable elliptic equations. SIAM J. Numer. Anal., 10, 1103.

Cooley, J.W., Lewis, P.A.W., and Welch, P.D. (1970) The fast Fourier transform algorithm: programming considerations in the calculation of sine, cosine and Laplace transforms. J. Sound Vib. 12, 315.

Cooley, J.W., and Tukey, J.W. (1965) An algorithm for machine calculation of complex Fourier series. Math. Computation 19, 297.

Crank, J. and Nicolson, P. (1947) A practical method for numerical evaluation of solutions of partial differential equations of the heat-conduction type. Proc. Camb. Phil. Soc. 43, 50.

Danielson, G.C., and Lanczos, C. (1942) Some improvements in practical Fourier analysis and their application to X-ray scattering from liquids. J. Franklin Inst. 233, 365.

Dennis, S.C.R. (1960) Finite differences associated with second-order differential equations. Q.J. Mech. Appl. Math. 13, 487.

Dobes, K. (1982) Algorithm 49, Fast Fourier Transforms with Recursively Generated Trigonometric Functions, Computing, 29 263.

Dorr, F.W. (1970) The direct solution of the discrete Poisson equation on a rectangle. SIAM Rev. 12, 248.

Dryja, M. (1982) A capacitance matrix method for Dirichlet problem on polygon region. Numer. Math. 39, 51.

D'Yakonov, E.G. (1961) On an iterative method for the solution of finite difference equations. Dokl. Aka. Nauk SSSR, 138, 522.

Eastwood, J.W. (1975) Optimal particle-mesh algorithms. J. Comput. Phys. 18, 1.

Evans, D.J. (1971) The numerical solution of the fourth boundary value problem for parabolic partial differential equations. J. Inst. Maths. Applics. 7, 61.

Evans, D.J. (1972) An algorithm for the solution of certain tridiagonal systems of linear equations. Comput. J. 15, 356.

Evans, D.J., and Atkinson, L.V. (1970) An algorithm for the solution of general three term linear systems. Comput. J. 13, 323.

Fornberg, B. (1981) A vector implementation of the fast Fourier transform. Math. Comput. 36, 189.

Gentleman, W.M., and Sande, G. (1966) Fast Fourier transforms - for fun and profit. Proceedings of AFIPS Fall Joint Computer Conf., Washington D.C., Spartan, 29, 563.

Gonzales, R.C., and Wintz, P. (1977) Digital Image Processing, Addison-Wesley, Reading, Massachusetts, U.S.A.

Good, I.J. (1958) The interaction algorithm and practical Fourier series. J. Roy. Stat. Soc. Ser. B. 20, 361.

Gottlieb, D., and Orszag, S.A. (1977) Numerical analysis of spectral methods. SIAM Regional Conference series in Applied mathematics.

Gunn, J.E. (1964) The numerical solution of $\nabla.(a\nabla u) = f$ by a semi-explicit alternating direction iterative method. Numer. Math. 6, 181.

Gunn, J.E. (1965) The solution of difference equations by a semi-explicit iterative technique. SIAM J. Numer. Anal. 2, 24.

Hildebrand, F.B. (1968) Finite-difference equations and simulations, Prentice-Hall, Englewood Cliffs, New Jersey.

Hockney, R.W. (1965) A fast direct solution of Poisson's equation using Fourier analysis. J. Ass. Comp. Mach. 12, 95.

Hockney, R.W. (1968) Formation and stability of virtual electrodes in a cylinder. J. Appl. Phys. 39, 4166.

Hockney, R.W. (1970) The potential calculation and some applications, Methods in Computational Physics 9, Academic Press, New York.

Hockney, R.W. (1978a) Computers Compilers and Poisson solvers, in Fast Poisson Solvers, edited by U. Schumann, Advance Publications Ltd, London.

Hockney, R.W. (1978b) POT4 - a FACR(1) algorithm for arbitrary regions, in Fast Poisson Solvers, edited by U. Schumann, Advance Publications Ltd, London.

Hockney, R.W., Warriner, R.A., and Reisser, M. (1974) Two-dimensional particle models in semi-conductor device analysis. Electronics Letters, 20, 484.

Householder, A.S. (1964) The theory of matrices in numerical analysis, Blaisdell, New York.

Houstis, E.N. and Papatheodorou, T.S. (1977) Comparison of fast direct methods for elliptic problems, in Advances in Computer methods for partial differential equations II, edited by R. Vichnevetsky, IMACS, New Brunswick, New Jersey, 46.

Huntley, E., Pickering, W.M., and Zinober, A.S.I. (1978) The
    numerical solution of linear time-dependent partial
    differential equations by the Laplace and fast Fourier
    transforms. J. Comput. Phys. 27, 256.

Hughes, M.H. (1971) Solution of Poisson's equation in cylindrical
    coordinates, Comput. Phys. Commun. 2, 157.

James, R.A. (1977) The solution of Poisson's equation for
    isolated source distributions. J. Comp. Phys. 25, 71.

Korn, D.L. and Lambiotte, J.J. (1979) Computing the fast Fourier
    transform on a vector computer. Math. Comput. 33, 977.

Le Bail, R.C. (1972) Use of fast Fourier transforms for solving
    partial differential equations in physics. J. Comput. Phys.
    9, 440.

Lorenz, E.N. (1976) A rapid procedure for inverting del-square
    with certain computers. Mon. Weather Rev. 104, 961.

Martin, E.D., (1974) A generalised capacity matrix technique for
    computing aerodynamic flows. Computers and Fluids 2, 79.

Matte, J-P. and Lafrance, G. (1977) Solution of the discrete
    Poisson equation with complicated boundaries. J. Comput.
    Phys. 23, 86.

Menikoff, R. and Zemach, C. (1980) Methods for numerical conformal
    mapping. J. Comput. Phys. 36, 366.

Mobley, C.D. and Stewart, R.J. (1980) On the numerical
    generation of boundary-fitted orthogonal curvilinear
    coordinate systems. J. Comput. Phys. 34, 124.

O'Leary, D.P. and Widlund, O. (1979) Capacitance matrix methods for the Helmholtz equation on general three-dimensional regions. Math. Comput. 33, 849.

Pickering, W.M. (1977) Some comments on the solution of Poisson's equation using Bickley's formula and fast Fourier transforms. J. Inst. Maths. Applics. 19, 337.

Pickering, W.M. (1983) Nonpolynomial finite difference schemes and the use of the fast Fourier transform. J. Comput. Phys. 51, 357.

Pickering, W.M. (1984). On the solution of a quasi-tridiagonal system of linear equations. Int. J. Comput. Math. 15, 181.

Pickering, W.M. (1986) On the solution of Poisson's equation on a regular hexagonal grid using FFT methods. J. Comput. Phys. to appear.

Pickering, W.M. and Sozou, C. (1979) The round laminar jet in a spherical envelope. J. Fluid Mech. 94, 1.

Proskurowski, W. (1979) Numerical solution of Helmholtz's equation by implicit capacitance matrix methods. ACM Trans. Math. Software 5, 36.

Proskurowski, W. and Widlund, O. (1976) On the numerical solution of Helmholtz's equation by the capacity matrix method. Math. Computation 30, 433.

Proskurowski, W. (1978) On the numerical solution of the eigenvalue problem of the Laplace operator by a capacitance matrix method. Computing 20, 139.

Proskurowski, W. and Widlund, O. (1980) A finite element capacity matrix method for the Neumann problem for Laplace's equation. SIAM J. Sci. Statist. Comput. $\underset{\sim}{1}$, 410.

Roscoe, D.F. (1975) New methods for the derivation of stable difference representations for differential equations. J. Inst. Maths. Applics, $\underset{\sim}{16}$, 291.

Singhal, K. and Vlach, J. (1975) Computation of time domain response by numerical inversion of the Laplace transform. J. Franklin Inst, $\underset{\sim}{299}$, 109.

Singhal, K. and Vlach, J. (1971) Program for numerical inversion of the Laplace Transform. Elec. Lett. $\underset{\sim}{7}$, 413.

Singhal, K., Vlach, J. and Nakhla, M. (1976) Absolutely stable, high order method for time domain solution of networks. Arch. Elek. Ubertr. $\underset{\sim}{30}$, 157.

Smith, G.D. (1978) Numerical solution of partial differential equations: Finite difference methods. Oxford University Press.

Smith, J. (1968) The coupled equation approach to the numerical solution of the biharmonic equation by finite differences I, SIAM J. Numer. Anal. $\underset{\sim}{5}$, 323.

Smith, J. (1970) The coupled equation approach to the numerical solution of the biharmonic equation by finite differences II, SIAM J. Numer. Anal. $\underset{\sim}{7}$, 104.

Sweet, R.A. (1973) Direct methods for the solution of Poisson's equation on a staggered grid. J. Comput. Phys. $\underset{\sim}{12}$, 422.

Sweet, R.A. (1977) A cyclic reduction algorithm for solving block tridiagonal systems of arbitrary dimension. SIAM J. Numer. Anal. 14, 706.

Swartztrauber, P.N. (1974a) A direct method for the discrete solution of separable elliptic equations. SIAM J. Numer. Anal. 11, 1136.

Swartztrauber, P.N. (1974b) The direct solution of the discrete Poisson equation on the surface of a sphere. J. Comput. Phys. 15, 46.

Swartztrauber, P.N. (1977) The methods of cyclic reduction, Fourier analysis and the FACR algorithm for the discrete solution of Poisson's equation on a rectangle. SIAM Rev. 19, 490.

Tang, I.C. (1969) A simple algorithm for solving linear equations of a certain type. Z.A.M.M. 49, 508.

Temperton, C. (1975) Algorithms for the solution of cyclic tri-diagonal systems. J. Comput. Phys. 19, 317.

Temperton, C. (1978) A fast Poisson-solver for an octagonal domain, in Computers, Fast Elliptic Solvers and Applications, edited by U. Schumann, Advance Publications Ltd., London.

Temperton, C. (1979) Direct methods for the solution of the discrete Poisson equation: some comparisons. J. Comput. Phys. 31, 1.

Temperton, C. (1980) On the FACR(ℓ) algorithm for the discrete Poisson equation. J. Comput. Phys. 34, 314.

Temperton, C. (1983a) Self-sorting mixed-radix fast Fourier transforms. J. Comp. Phys. 52, 1.

Temperton, C. (1983b) Fast mixed-radix real Fourier transforms. J. Comp. Phys. 52, 340.

Temperton, C. (1983c) A note on prime factor FFT algorithms. J. Comp. Phys. 52, 198.

Thom, A. and Apelt, C.J. (1961) Field Computations in Engineering and Physics. Van Nostrand Reinhold, London.

Thomas, L.H. (1949) Elliptic problems in linear difference equations over a network. Watson Sci. Comput. Lab. Rept. Columbia University, New York.

Varga, R.S. (1962) Matrix iterative analysis. Prentice-Hall, Englewood Cliffs, New Jersey.

Vlach, J. (1969) Numerical method for transient responses of linear networks with lumped, distributed or mixed parameters J. Franklin. Inst., 288, 99.

Whittaker and Robinson (1944) Calculus of observations, Blackie and Son, London.

Widlund, O. (1972) On the use of fast methods for separable finite-difference equations for the solution of general elliptic problems, in Sparse Matrices and their Applications, edited by D.J. Rose and R.A. Willoughby, Plenum Press, New York.

Wilhelmson, R.B., and Ericksen, J.H. (1977) Direct solutions for Poisson's equation in three dimensions. J. Comput. Phys. 25, 319.

Wilkinson, J.H. (1961) Error analysis in direct methods of matrix
inversion. J. Ass. Comput. Mach. 8, 281.

Williams, G.P. (1969) Numerical integration of the three-
dimensional Navier-Stokes equations for incompressible flow.
J. Fluid. Mech. 37, 727.

Index

ADI method,
  see Peacemann-Rachford
Amplification factor, 99
Amplification matrix, see Matrix
Arithmetic operations,
  capacitance matrix, 120,125
  cosine transform, 45
  FFT algorithm, 10,30
  Poisson eqn., 6,20,83,105,106
  sine transform, 47
  tridiagonal algorithms, 15,17

Biharmonic equation, 74,133
Binary digits, 24,27,30
Bit-reversed, 25,29,31
Block elimination, 89
Boundary conditions, 7,51,56
  Dirichlet, 2,53,84,146,154
  Dirichlet-Neumann, 14,159
  Neumann, 7,14,15,54,81,155
  Periodic, 7,14,15,54,157
  time-varying, 151
Bunemann variant 1,101,132
         variant 2,104
  see also Cyclic reduction

Capacitance matrix, 113,119,
  120,123,131,133
  circulant, 121
  finite element calc., 134
  inversion of, 121,125,133
  eigenproblems, 134
Chebyshev acceleration, 137
  parameters for, 139

Cholesky method, 120
Concus & Golub iteration, 136
  parameter for, 138
Conformal mapping, 111
Cosine transform, 14,42,81,155
Crank-Nicolson scheme, 67
Cyclic reduction, 7,18,85,147
  block, 90
  Bunemann variant 1,101,132
          variant 2,104
  CORF algorithm, 97
  eliminated eqns., 89,97
  generalisations, 105
  reduced eqns., 89
  stability, 98

Data sequence, see Sequence
DFT, IDFT, 8,9,23
  complex data, 36,38
  cosine transform, 14,42,81,155
  real data, 36,40,41,157
  sine transform, 13,45,154
  sine transform, shifted, 159
Diagonal dominance, 15,61,63
Discrete representations, 51
  see also Finite difference
Dual-node computation, 28
D'Yakonov-Gunn iter., 136,142

Eigenvalues, vectors, 153
  Dirichlet problem, 4,5,63,68,71,
    84,154
  Dirichlet-Neumann problem, 159
  Neumann problem, 81,82,155

176